Lecture Notes in Mathematics

Edited by A. Dold and B. Eckmann

1249

Imme van den Berg

Nonstandard Asymptotic Analysis

Springer-Verlag
Berlin Heidelberg New York London Paris Tokyo

Author

Imme van den Berg
Institute of Econometrics, Faculty of Economics
P.O. Box 800, 9700 AV Groningen, Holland

Mathematics Subject Classification (1980): 03 E 15, 03 H 10, 06 F 15, 40 A 25, 41 A 60, 65 B 10

ISBN 3-540-17767-1 Springer-Verlag Berlin Heidelberg New York
ISBN 0-387-17767-1 Springer-Verlag New York Berlin Heidelberg

© Springer-Verlag Berlin Heidelberg 1987
Printed in Germany

Printing and binding: Druckhaus Beltz, Hemsbach/Bergstr.
2146/3140-543210

ACKNOWLEDGEMENTS.

This book is the fruit of an intensive cooperation with E. Benoit, J.-L. Callot and F. and M. Diener, during a stay in Algeria. I was deeply influenced by Professor R. Lutz and Professor G. Reeb, who also gave invaluable advise in the preparation of this manuscript. This text further benefited from discussions with among others M. Goze, J. Harthong and J.-P. Reveilles. All the above mathematicians speak the Alsatian dialect of nonstandard analysis.

I am greately indebted to Professor E.M. de Jager (Amsterdam) and Professor J.-P. Ramis (Strasbourg) who drew my attention to asymptotic theory. The latter insisted on the problem of the "summation to the smallest term" which plays a central role in this book.

I would like to express special thanks to the Institute of Econometrics of the University of Groningen, which gave me the opportunity to finish the manuscript, and to Mrs. A. van Oosten, who typed it with such great savoir-faire.

INTRODUCTION.

1). This book has three main purposes, firstly to present nonstandard methods of asymptotic reasoning, secondly to present new results obtained by them, and thirdly to present a nonstandard alternative to the classical theory of asymptotic expansions. It is adressed to mathematicians knowing the basic principles of nonstandard analysis, who are willing to learn asymptotics and to classical asymptoticists, who wish to form an opinion about the relevancy of nonstandard analysis to this branch of mathematics; the latter will observe that problems very familiar to them are treated in a manner clearly distinct from the classical one, and may compare the efficiency of both approaches.

2). It is common practice among asymptoticists to speak informally about "fixed" numbers, to be distinguished from numbers depending on a "large" or "small" parameter. Nonstandard analysis legalizes this matter of speaking. A formal predicate "standard" is introduced, and it is postulated that there exist "unlimited" real numbers larger than any standard real number, and "infinitesimals", smaller in absolute value than any strictly positive standard real number.

Once the informal predicate "fixed" is transformed into a formal predicate "standard" we can define collections with it and operate on them. Normally these collections fall outside classical mathematics. Then they are called "external sets". Thus we dispose of different orders of magnitude, domains of asymptotic behaviour of functions and domains of validity of asymptotic reasonings, all within $\mathbb{R}$. What is more, we can define external relations, a valuable tool in two-parameter problems, formulate general lemma's on infinitesimal approximation, and derive permanence principles. The latter extend the validity of a property proved by direct means on some domain to a larger domain; they are widely applicable and enable us to match two asymptotic expressions of a function, established on neighbouring regions. Together the methods based on external sets are a powerful tool in solving asymptotic problems.

The new results in asymptotics we derived with them concern the remainders of convergent and divergent Taylor series and asymptotic series. We established asymptotic expressions of the remainders for a large class of functions and used these expressions to study two problems.

The first problem is a very old one (Euler, Poincaré, Stieltjes, Nevanlinna....): how to use divergent series in approximations? It is well-known from practice that the use of divergent series yields remarkable approximation results, often defying convergent expansions: using only a few terms a considerable precision is attained. However, numerical practice also shows that divergent series should be handled with care. Indeed, with increasing index the speed of the approximation slows down, until some optimal remainder is attained. Then no more terms should be added, for the approximation will get worse: at first slowly, but the more rapidly as the index grows large. Using nonstandard analysis it is possible to give a complete, asymptotic description of this process. In particular we were able to prove for an important class of functions another well-known fact of numerical practice: the index of the smallest term of divergent expansions approximately coincides with the index of best approximation.

The second problem is also old, and it has become wider known since the use of microcomputers: it appears that the configuration of the Taylor polynomials of some familiar functions, like the exponential or the sine function is strikingly regular. We propose a mathematical description of this phenomenon, based on the asymptotic expressions of the remainders, and an entirely nonstandard, external way of modelling visual observations into mathematical language.

The nonstandard alternative to the theory of asymptotic expansions of functions, which dates back to Poincaré (1886), are the shadow-expansions of numbers. It is my conviction that the latter fit better to the numerical observations on divergent series mentioned above. If a number possesses a shadow expansion it immediately follows from the definition that for not too large indices the remainders decrease rapidly; this property is not a trivial consequence of the theory of asymptotic expansions of functions. Of course I do not advocate to replace the theory of asymptotic expansions, which is well established and has been very succesful, by a theory of shadow expansions, which is only in construction. However, the latter has recently found applications in nonstandard perturbation theory (See the work of F. and M. Diener, Benoit, Callot and Reeb on the "canards" and the "streams").

3). (about nonstandard analysis). This study is presented within an axiomatic form of nonstandard analysis, namely Internal Set Theory (IST), founded by Nelson. This book does not contain an introduction to IST. There are many introductions available now: a list of them is found among the references. Though it has a strong intuitive flavor the reader without much experience in nonstandard analysis may have some difficulty with nonstandard terminology. For this reason we joined a lexicon at the end of this book.

4). (historical note). We mention briefly three attempts to develop further or modify Poincaré's approach to asymptotic analysis. The first and second are by Van der Corput. The reader with a nonstandard background will not fail to recognize lines of thought which bear a definite resemblance to nonstandard analysis. In "Asymptotic developments I: Fundamental theorems of asymptotics" (1956) Van der Corput, after a short introduction adopts "an important convention: each number and each function which occurs in pure asymptotics and which is not called fixed, may depend on ω, whereas fixed means independent of ω". He does not refrain from quantifying over fixed numbers. So he defines "asymptotically finite numbers" to be numbers A such that there are fixed $c,p,\gamma > 0$ such that $|A| \leqq c\omega^p$ if $\omega \geqq \gamma$. "Asymptotically equal" are two numbers B and C such that for all fixed $q > 0$ there exist fixed $c,\gamma > 0$ such that $|B-C| \leqq c\omega^{-q}$ if $\omega \geqq \gamma$. He introduces a notion of "asymptotic limit" which contains up to three changes of quantifiers over "fixed numbers". Furthermore he develops an important algebra concerning these notions.

In "Introduction to the neutrix calculus" (1961) he introduces so-called "neutrices", sets of functions which are "neglectable" in some specific asymptotic problem. Such a neutrix may be the set of real functions f(x) tending to 0 if x tends to infinity. To get rid of the neglectable functions in the asymptotic problem he quotients the set of real functions by the neutrix. He treats the resulting classes much like real numbers. This approach is not unlike A. Robinson's construction of the nonstandard model of the real numbers $^*\mathbb{R}$: two real functions belong to the same nonstandard real number (=class of real functions) if they are equal on a neighbourhood of infinity. Van der Corput even disposes of a "principle of permanence" which acts as a transfer principle.

The third attempt is within nonstandard analysis. In "Nonarchimedean fields and asymptotic expansions" (1976) A.H. Lightstone and A. Robinson define a nonarchimedean field $^\rho\mathbb{R}$ situated between the standard model $\mathbb{R}$ and the nonstandard model $^*\mathbb{R}$. If we replace "fixed" by "standard" and take ω unlimited in Van der Corput's definition, the field $^\rho\mathbb{R}$ is nothing else than the set of asymptotically finite numbers quotiented by the numbers which are asymptotically equal to 0. The set $^\rho\mathbb{R}$ serves as a starting point for an asymptotic theory of numbers, but the authors do not pursue their research in this direction very far.

5). (structure of this work). This book is divided into three parts. In the first part (Chapter I) we treat in a nonstandard way very familiar concrete examples of asymptotics: the exponential integral, Stirling's formula, the divergent expansions of the Bessel functions, the Taylor polynomials of the exponential function. We conclude with a list of the main nonstandard methods we used. It was our intention to proceed very didactically in this introductory chapter. We hope it can be read by anyone who understands the nonstandard structure of the real line: the infinitesimals in the middle, around 0, surrounded by the limited numbers, at the left end the negative unlimited numbers and at the right end the positive unlimited numbers.

In the second part (Chapter II and III) we embed the results obtained on the examples into a general theory. In Chapter II we derive under fairly general conditions on the coefficients asymptotic expressions for the remainders of both convergent and divergent expansions. They enable a rather precise qualitative description of the approximation properties of these expansions. A still more precise description is given in Chapter III, valid in the case some additional conditions are satisfied (regular behaviour of coefficients, growth conditions......)

The second part requires more understanding of basic principles of nonstandard analysis than the first part, like Transfer, S-continuity, shadow and permanence principles.

In the third part (Chapter IV, V and VI) we develop in a linear way the foundations of a general theory of nonstandard asymptotics. In particular the Chapters IV and VI use a substantial part of Internal Set Theory. Chapter IV contains the theory of external sets. The reader rightly wonders whether it is necessary to study set theory in order to understand nonstandard asymptotics. Indeed, it happens that in studying concrete examples one can do without, at the price of introducing technical arguments. However, the investment in procedures based on external sets is largely compensated by a greater ease in handling asymptotic problems, and a gain in generality. In Chapter V we present general approximation lemma's, used throughout this book. The most important is the lemma of dominated approximation, which resembles Lebesgue's dominated convergence theorem. In Chapter VI the foundations are laid for a theory of shadow expansions. Special attention is paid to the Goze decomposition of an infinitesimal vector. Unlike Taylor, asymptotic or shadow expansions, this decomposition is defined without any restrictions. If some conditions are satisfied, this decomposition yields a shadow expansion.

Each chapter, and most sections, opens with an introduction, which contains more detailed information about their contents.

Many tables and diagrams illustrate the numerical relevancy of the asymptotic results. They were obtained on a HP 75 C microcomputer.

CONTENTS

page

PART III

PART I

CHAPTER I. FOUR EXAMPLES OF NONSTANDARD REASONING IN ASYMPTOTICS

In this chapter methods and techniques, which will be developed systematically in part III are applied to some elementary examples of asymptotics. With this presentation, the reader who is familiar with asymptotic approximations, but has no experience with nonstandard analysis, may get some feeling about nonstandard methods; also the reader who is aquainted with nonstandard analysis, but not with asymptotic approximations, may get some idea of asymptotics. Furthermore, the elementary setting may be helpful in comparing the efficiency of nonstandard and classical methods.

We consider

1. The behaviour of the remainder $R_n(\varepsilon)$ associated to the divergent expansion $\sum_{k=0}(-1)^k k!\,\varepsilon^k$ of the integral $\int_0^\infty \frac{e^{-t}}{1+\varepsilon t}dt$ as a function of n, where ε is a fixed positive infinitesimal number.
2. Stirling's formula.
3. The divergent development at infinity of the Besselfunctions $J_p(z)$.
4. The behaviour of the remainder associated to the Taylorexpansion of e^x, as a function of x.

For very basic nonstandard notions, like "infinitesimal" we refer to the lexicon. Otherwise, whenever a nonstandard tool is met for the first time, a precise reference is stated. Many nonstandard notions have a strong intuitive sense, and we intend to proceed very didactically. However, readers not acquainted with nonstandard analysis may find it profitable to accompany the reading of parts of this chapter with some introductory paper. A list of introductory papers is found at the end of this book.

1. The approximation of the number $\int_0^\infty \frac{e^{-t}}{1+\varepsilon t}dt$ by the partial sums of the expansion $\sum_{k=0}(-1)^k k!\,\varepsilon^k$.

§1. Review and notation.

We define the function f for $x \geqq 0$ by

$$f(x) = \int_0^\infty \frac{e^{-t}}{1+xt}dt$$

The function f is related to the well-known exponential integral E_1 defined by

$$E_1(x) = \int_1^\infty \frac{e^{-xt}}{t}dt \qquad \left(= \frac{e^{-x}}{x}f(1/x)\right)$$

The kth term of the Taylor-series of the function f equals $(-1)^k k!x^k$. To express

n	$R_n(\varepsilon)$	$R_n(\varepsilon)/T_{n+1}(\varepsilon)$	$R_{n+1}(\varepsilon)/R_n(\varepsilon)$
0	-9.980059761E-004	0.9980059761	-0.0019980080
1	1.994023881E-006	0.9970119404	-0.0029970149
2	-5.976119285E-009	0.9960198808	-0.0039960238
3	2.388071500E-011	0.9950297917	-0.0049950347
4	-1.192850000E-013	0.9940416670	-0.0059940475
5	7.149999607E-016	0.9930555010	-0.0069930623
6	-5.000039291E-018	0.9920712878	-0.0079920791
7	3.996070936E-020	0.9910890217	-0.0089910977
8	-3.592906439E-022	0.9901086969	-0.0099901184
9	3.589356060E-024	0.9891303077	-0.0109891409
10	-3.944393953E-026	0.9881538482	-0.0119881654
11	4.728604704E-028	0.9871793130	-0.0129871918
12	-6.141129610E-030	0.9862066961	-0.0139862201
13	8.589119022E-032	0.9852359921	-0.0149852503
14	-1.287100982E-033	0.9842671952	-0.0159842824
15	2.057338557E-035	0.9833002999	-0.0169833164
16	-3.494043166E-037	0.9823353006	-0.0179823523
17	6.283111515E-039	0.9813721916	-0.0189813901
18	-1.192621906E-040	0.9804109675	-0.0199804297
19	2.382909820E-042	0.9794516228	-0.0209794713
20	-4.999218813E-044	0.9784941519	-0.0219785147

Figure 1. Approximation of the number $\int_0^\infty \frac{e^{-t}}{1+\varepsilon t}dt$ by partial sums $\sum_{k=0}^{n}(-1)^k k!\varepsilon^k$ of its divergent series-expansion; $\varepsilon = 0.001$. When the degree n is not too large the remainders $R_n(\varepsilon)$ decrease fairly rapidly (column 2) and are closely related to the first neglected term (column 3). The relative amelioration of the approximation result $R_{n+1}(\varepsilon)/R_n(\varepsilon)$ depends in an almost linear way on the index n+2 (column 4)[(1)].

the divergence of the Taylorseries we write

$$f(x) \sim \sum_{k=0}(-1)^k k!x^k$$

The expansion may be derived in several ways: firstly by direct differentiation and secondly by repeated integration by parts. A third way is to expand $\frac{1}{1+xt}$ in the power series $\sum_{k=0}^{\infty}(-1)^k(tx)^k$, and then integrate term-by-term. As it turns out, the method of repeated integration by parts provides the simplest formula for the remainder. We have

$$f(x) = \sum_{k=0}^{n-1}(-1)^k k!x^k + (-1)^n n!x^n \int_0^\infty \frac{e^{-t}}{(1+xt)^{n+1}}dt$$

Let us write

$$T_n(x) = (-1)^n n!x^n$$

$$S_n(x) = \sum_{k=0}^{n} T_k(x)$$

$$R_n(x) = f(x) - S_n(x)$$

(1). The symbol "E" means "times 10 to the power ". So E-004=.10^{-4} etc.

The symbol $\emptyset$: In classical asymptotics one uses the expression '$\varphi(x) = o(1)$ for $x \to 0$' in case φ is a function which tends to zero when the variable x tends to zero. It is used when the exact knowledge of the function φ is of no importance, and only the behaviour of φ in the vicinity of 0, i.e. its tendency towards 0 matters. Here we introduce the symbol $\emptyset$, the zero of the computers, which is to be used in the same spirit.

The symbol $\emptyset$ designates an infinitesimal number. It is used when exact knowledge of the number is of no importance, but only its order of magnitude (namely to be infinitesimal). Corresponding to the use of the symbol $o(1)$, the symbol $\emptyset$ may occur more than once in one expression, while designating different infinitesimals. It is also admitted that the infinitesimals depend on indices.

§2. Formulation of the problem and principal results.

In spite of the divergence of the series $\sum_{k=0}^{\infty} (-1)^k k!\varepsilon^k$ the numbers $f(\varepsilon)$ are very well approximated by its partial sums $S_n(\varepsilon)$, for small ε. Indeed, the 2nd and 4th columns of Fig. 1 show that for $\varepsilon = 0.001$ the remainder $R_n(\varepsilon)$ decreases rapidly when n increases from 0 to 20. We may hope to explain this phenomenon by relating the remainder to the first neglected term of the expansion. Indeed, (Fig. 1, 3rd column), again for $\varepsilon = 0.001$,

the remainder and the first neglected term are almost equal. Of course, not all sums $S_n(\varepsilon)$ furnish a good approximation of $f(\varepsilon)$. This is already illustrated by the last column of Fig. 1, which shows that the relative improvement resulting from the addition of a new term slows down with increasing n. In fact, the remainders diminish gradually towards a minimum, and then start to increase, as shown by Figures 2 and 3. These figures also show that the minimum is attained for some index, which is close to that of the smallest term ($[1/\varepsilon]$).

We will try to give a mathematical description of the observed numerical phenomena. The data are approximative, so this suggests a qualitative, or a nearly quantitative approach. Such an approach is further motivated by the remarkable regularity which persists throughout the figures (see also Fig. 4).

In order to describe the phenomena observed on the Figures 1 and 2 we must take ε fixed. Notice that for $\varepsilon = 0.1$ the phenomena are still observed, but are less precise than for $\varepsilon = 0.01$ or $\varepsilon = 0.001$. For ε of order 1 the phenomena disappear (Fig. 5). So we must take ε small. We choose to consider the ideal situation where ε is infinitesimal. As we will show, it is then not only possible to give a qualitative description of the observed phenomena, but even a nearly-quantitative, or asymptotic description. Here are the main results.

Proposition 1.1. For all $n \in \mathbb{N}$ we have $R_{n-1}(\varepsilon) = (1+\emptyset)\dfrac{T_n(\varepsilon)}{1+(n+1)\varepsilon}$.

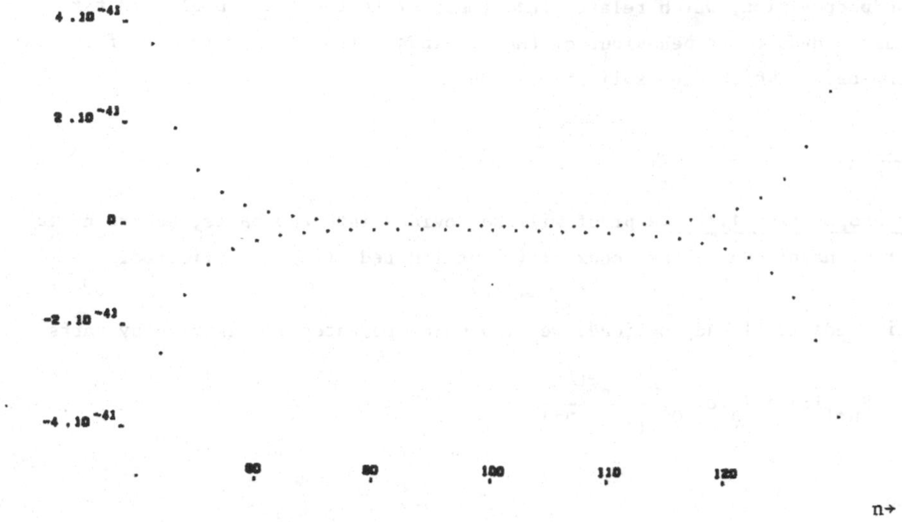

Figure 2. The remainder $R_n(\varepsilon)$ as a function of n for $\varepsilon = 0.01$. Optimal approximation result is attained for indices close to the index of the smallest term $((-1)^n n!\varepsilon^n)$. Here the index of the smallest term equals 99 or 100.

Proposition 1.2. For all $n \in \mathbb{N}$ we have $R_n(\varepsilon)/R_{n-1}(\varepsilon) = (1+\emptyset)T_{n+1}(\varepsilon)/T_n(\varepsilon) =$

$$= -(1+\emptyset)(n+1)\varepsilon$$

Proposition 1.3. If $|R_\omega(\varepsilon)|$ is minimal, then ω is of the form $\omega = 1/\varepsilon+\emptyset/\sqrt{\varepsilon}$. For all indices ω of the form $1/\varepsilon+\emptyset/\sqrt{\varepsilon}$ we have

$$|R_\omega(\varepsilon)| = (1+\emptyset)\sqrt{\frac{\pi}{2\varepsilon}}e^{-1/\varepsilon}$$

The first proposition, which relates the remainder to the first neglected term, enables us to deduce the behaviour of the remainders from the behaviour of the terms of the expansion, which are explicitly known.

§3. Proofs.

Proof of proposition 1.1. The proof will be devided into two parts, according to the order of nagnitude of the index n: (i) $n\varepsilon$ limited, (ii) $n\varepsilon$ unlimited.

(i) $n\varepsilon$ limited: As already noticed, we find using repeated integration by parts

$$R_{n-1}(\varepsilon) = T_n(\varepsilon) \int_0^\infty \frac{e^{-t}}{(1+\varepsilon t)^{n+1}} dt$$

We define

$$I(t) = \frac{e^{-t}}{(1+\varepsilon t)^{n+1}}$$

We must approximate $\int_0^\infty I(t)dt$. At this end we will use an important lemma of non-standard asymptotics, namely the lemma of dominated approximation (in integral form, lemma 5.2). According to the hypotheses of this lemma, we have to approximate $I(t)$ for limited t by a function, say $J(t)$, and majorize $I(t)$ (and $J(t)$) for all t by a standard function. This suffices to conclude that $\int_0^\infty I(t)dt \simeq \int_0^\infty J(t)dt$.
In very simple manner we find

$$I(t) \simeq e^{-(1+(n+1)\varepsilon)t} \qquad \text{(t limited)}$$

$$I(t), e^{-(1+(n+1)\varepsilon)t} \leqq e^{-t} \qquad \text{(all } t \geqq 0)$$

So

$$\int_0^\infty I(t)dt \simeq \int_0^\infty e^{-(1+(n+1)\varepsilon)t} dt = \frac{1}{1+(n+1)\varepsilon}$$

Hence $R_{n-1}(\varepsilon) = (1+\emptyset)\dfrac{T_n(\varepsilon)}{1+(n+1)\varepsilon}$.

(ii) $n\varepsilon$ unlimited: We write $p = [3/\varepsilon]$ (the choice of the number 3 is somewhat arbitrary, any standard number larger than 1 would do). Using part (i) of the proof we get

$$R_{n-1}(\varepsilon) = R_{p-1}(\varepsilon) - \sum_{k=p}^{n-1} T_k(\varepsilon) = (1+\emptyset)\frac{T_p(\varepsilon)}{4} - T_{n-1}(\varepsilon) \sum_{m=1}^{n-p} T_{n-m}(\varepsilon)/T_{n-1}(\varepsilon).$$

n	$S_n(\varepsilon)$	$R_n(\varepsilon)$	$\frac{R_n(\varepsilon)}{T_{n+1}(\varepsilon)}$	$\frac{R_{n+1}(\varepsilon)}{R_n(\varepsilon)}$
0	1.0000000000	-0.0843666606	0.844	-0.185
1	0.9000000000	0.0156333394	0.782	-0.279
2	0.9200000000	-0.0043666606	0.728	-0.374
3	0.9140000000	0.0016333394	0.681	-0.469
4	0.9164000000	-0.0007666606	0.639	-0.565
5	0.9152000000	0.0004333394	0.602	-0.662
6	0.9159200000	-0.0002866606	0.569	-0.758
7	0.9154160000	0.0002173394	0.539	-0.855
8	0.9158192000	-0.0001858606	0.512	-0.952
9	0.9154563200	0.0001770194	0.488	-1.050
10	0.9158192000	-0.0001858606	0.466	-1.148
11	0.9154200320	0.0002133074	0.445	-1.246
12	0.9158990336	-0.0002656942	0.427	-1.344
13	0.9152763315	0.0003570079	0.410	-1.442
14	0.9161481144	-0.0005147750	0.394	-1.540
15	0.9148404401	0.0007928993	0.379	-1.639
16	0.9169327191	-0.0012993797	0.365	-1.737
17	0.9133758448	0.0022574946	0.353	-1.836
18	0.9197782185	-0.0041448791	0.341	-1.935
19	0.9076137084	0.0080196310	0.330	-2.034
20	0.9319427285	-0.0163093891	0.319	-2.133

Figure 3. Approximation of the number $\int_0^\infty \frac{e^{-t}}{1+\varepsilon t}dt = 0.9156333394\ldots$ by partial sums $S_n(\varepsilon)$ of its divergent series expansion, $\varepsilon = 0.1$. The remainders first decrease, attain an optimum ($n=9$, the first index for which the term is minimal), then increase. The behaviour of the approximation still verifies grosso modo the qualitative description of §4 (below), however, with less precision than in the case $\varepsilon = 0.001$ (Fig. 1.).

We define $T_{n-m}(\varepsilon)/T_{n-1}(\varepsilon) = U_m$ and approximate $\sum_{m=1}^{n-p} U_m$ using again the lemma of dominated approximation (now in series form, lemma 5.11). We put $V_1=1$ and $V_m=0$ for $m \geq 2$. Now we have $U_m \simeq V_m$ for limited m, for

$$U_m = 1 \qquad m = 1$$

$$U_m = \frac{(-\varepsilon)^{1-m}}{(n-m+1)\ldots(n-1)} \simeq 0 \qquad m \geq 2$$

For all m we have the standard majoration

$$|U_m|, V_m \leq (1/2)^m$$

So by lemma 5.11

Figure 4.
Last column of Fig. 1, rounded off to 3 decimals, and last column of Fig. 3, rounded off to 1 decimal.

n	$R_{n+1}(\varepsilon)/R_n(\varepsilon)$ $\varepsilon=0.001$	$\varepsilon=0.1$
0	-0.002	-0.2
1	-0.003	-0.3
2	-0.004	-0.4
3	-0.005	-0.5
4	-0.006	-0.6
5	-0.007	-0.7
6	-0.008	-0.8
7	-0.009	-0.9
8	-0.010	-1.0
9	-0.011	-1.0
10	-0.012	-1.1
11	-0.013	-1.2
12	-0.014	-1.3
13	-0.015	-1.4
14	-0.016	-1.5
15	-0.017	-1.6
16	-0.018	-1.7
17	-0.019	-1.8
18	-0.020	-1.9
19	-0.021	-2.0
20	-0.022	-2.1

$$\sum_{m=1}^{n-p} U_m \simeq \sum_{m=1}^{\infty} V_m = 1$$

Hence

$$R_{n-1}(\varepsilon) = (1+\emptyset)\frac{T_p(\varepsilon)}{4} - (1+\emptyset)T_{n-1}(\varepsilon) = -(1+\emptyset)T_{n-1}(\varepsilon) = (1+\emptyset)\frac{T_n(\varepsilon)}{n\varepsilon} =$$

$$= (1+\emptyset)\frac{T_n(\varepsilon)}{1+(n+1)\varepsilon}.$$

Proof of proposition 1.2: The result is a consequence of proposition 1.1. Indeed,

$$\frac{R_n(\varepsilon)}{R_{n-1}(\varepsilon)} = \frac{(1+\emptyset)(1+(n+2)\varepsilon)T_{n+1}(\varepsilon)}{(1+\emptyset)(1+(n+1)\varepsilon)T_n(\varepsilon)} = (1+\emptyset)\frac{T_{n+1}(\varepsilon)}{T_n(\varepsilon)} = -(1+\emptyset)(n+1)\varepsilon$$

Proof of proposition 1.3: It follows from proposition 1.2 that $|R_n(\varepsilon)/R_{n-1}(\varepsilon)| < 1$ if $n\varepsilon \lnapprox 1$ and that $|R_n(\varepsilon)/R_{n-1}(\varepsilon)| > 1$ if $n\varepsilon \gnapprox 1$. So if $|R_\omega(\varepsilon)|$ is minimal, then ω must be of the form $\omega = \frac{1+\alpha}{\varepsilon}$ with $\alpha \simeq 0$. For such indices we get, using Stirling's formula (see section I.2)

$$|R_{\omega-1}(\varepsilon)| = (1+\emptyset)\frac{\omega!\ \varepsilon^{\omega}}{1+\varepsilon(\omega+1)}$$

$$= (1+\emptyset)\frac{\omega^{\omega}e^{-\omega}\sqrt{\omega}\sqrt{2\pi}\varepsilon^{\omega}}{2}$$

$$= (1+\emptyset)\sqrt{\frac{\pi}{2\varepsilon}}\ e^{\frac{(1+\alpha)}{\varepsilon}\log(1+\alpha)-1/\varepsilon-\alpha/\varepsilon}$$

$$= (1+\emptyset)\sqrt{\frac{\pi}{2\varepsilon}}\ e^{\frac{(1+\alpha)}{\varepsilon}(\alpha-\frac{\alpha^2}{2}(1+\emptyset))-1/\varepsilon-\alpha/\varepsilon}$$

$$= (1+\emptyset)\sqrt{\frac{\pi}{2\varepsilon}}\ e^{-1/\varepsilon+(1+\emptyset)\frac{\alpha^2}{2\varepsilon}}$$

So if ω is such that $R_{\omega}(\varepsilon)$ is minimal, the condition $\alpha^2/\varepsilon \simeq 0$ must be fulfilled. Hence ω has to be of the form $1/\varepsilon+\emptyset/\sqrt{\varepsilon}$. For such indices we have

$$|R_{\omega}(\varepsilon)| = (1+\emptyset)\sqrt{\frac{\pi}{2\varepsilon}}\ e^{-1/\varepsilon}$$

§4. Qualitative description of the approximation as a function of the degree.

With the help of the three propositions above we are able to describe the behaviour of the sequence of remainders as a function of n.

Generally speaking, the behaviour of the remainders is closely related to the behaviour of the terms. This follows from proposition 1.1 (the variation of the factor $\frac{1}{1+(n+1)\varepsilon}$ is almost negligible with respect to the rather violent variation of $T_n(\varepsilon)$) and from the first near-equality of proposition 1.2 ($R_n(\varepsilon)/R_{n-1}(\varepsilon) = (1+\emptyset)T_{n+1}(\varepsilon)/T_n(\varepsilon)$, valid for all $n \in \mathbb{N}$).

The formula of proposition 1.1 also shows that the connection between a new term of the expansion and the actual remainder, which is very strong in the beginning ($T_n(\varepsilon)/R_{n-1}(\varepsilon) \simeq 1$ as long as $n\varepsilon \simeq 0$, the new term almost wipes out the actual remainder), gets slowly but surely weaker and weaker. The terms become more and more excedentary and contribute less to the amelioration of the approximation.

The slow transition from amelioration to deterioration of the approximation is well described by the second near equality of proposition 1.2 ($R_n(\varepsilon)/R_{n-1}(\varepsilon) = -(1+\emptyset)(n+1)\varepsilon$, which implies that the quotient $|R_n(\varepsilon)/R_{n-1}(\varepsilon)|$ increases nearly linearly with n+1. (if n is very large, it is more precise to say "with n", for $R_n(\varepsilon)$ is then to be compared with $T_n(\varepsilon)$ instead of $T_{n+1}(\varepsilon)$; see Fig. 4).

The formula shows that $R_n(\varepsilon)$ decreases rapidly as long as $n\varepsilon \simeq 0$. There is question of sensible amelioration as long as $n\varepsilon \lessapprox 1$, but the relative amelioration slows down. For $n\varepsilon \simeq 1$ the partial sums oscillate around the exact value with amplitude nearly equal to half of the modulus of the terms. The approximation is nearly optimal for a proper subset of these indices, namely the indices whose

Figure 5:
The quotient $T_{n+1}(x)/R_n(x)$ for x of order 1 and low n. For $\varepsilon \simeq 0$ it is shown that $T_{n+1}(\varepsilon)/R_n(\varepsilon) = (1+\emptyset)(1+(n+2)\varepsilon)$. This formula is here no longer well observed. Notice that the terms mostly are highly excedentary as regards to the actual remainders, and that there still appears some regularity through the data.

n \ x	0.5	1.0	2.0
0	1.803	2.477	3.714
1	2.246	3.354	5.474
2	2.704	4.275	7.341
3	3.174	5.222	9.262
4	3.650	6.184	11.210
5	4.132	7.157	13.175

distance to $1/\varepsilon$ is small with respect to $1/\sqrt{\varepsilon}$; the remainder is then very small, of order $e^{-1/\varepsilon}$. For $n\varepsilon \gtrsim 1$ the approximation worsens, and the more rapidly as n is large.

§5. Comments on the results.

1. It is very easy to derive error bounds for the above expansion of the number $f(\varepsilon)$. Indeed, for all $\varepsilon > 0$

$$|R_{n-1}(\varepsilon)| = |(-1)^n n!\varepsilon^n \int_0^\infty \frac{e^{-t}}{(1+\varepsilon t)^{n+1}} dt| \leqq |n!\varepsilon^n \int_0^\infty e^{-t} dt| = |T_n(\varepsilon)|.$$

Hence, as usually remarked in textbooks, the index of the least error bound coincides with the index of the smallest term. Although it is plausible, of course this does not imply that the index of the least actual error coincides with the index of the smallest term.

Proposition 1.3 states that this is nearly true in the ideal case where ε is infinitesimal. More precisely, it states that the index $\omega = \omega(\varepsilon)$ of optimal approximation is of the form $1/\varepsilon+\emptyset/\sqrt{\varepsilon}$; so it is near the index of the smallest term, which equals $[1/\varepsilon]$.

2. How to formulate such a result in classical mathematics? We proved

(*) $\qquad (\forall \varepsilon \simeq 0)(\omega(\varepsilon)-1/\varepsilon = \emptyset/\sqrt{\varepsilon})$

We now get rid of the nonclassical symbols $\simeq$ and $\emptyset$. Let $s > 0$ be a standard number. The following proposition is a consequence of (*).

$$(\forall^{st}\ s > 0)(\forall \varepsilon \simeq 0)(|\omega(\varepsilon)-1/\varepsilon| < s/\sqrt{\varepsilon})$$

So we have

$$(**) \qquad (\forall^{st}\ s > 0)(\exists d > 0)(\forall \varepsilon \leqq d)(|\omega(\varepsilon)-1/\varepsilon| < s/\sqrt{\varepsilon})$$

Indeed, any infinitesimal d suffices. We now apply one of the axioms of Internal Set Theory. We notice that

$$(\exists d > 0)(\forall \varepsilon \leqq d)(|\omega(\varepsilon)-1/\varepsilon|) < s/\sqrt{\varepsilon})$$

has no free variables, except for s and ω, which are both standard. Then by the Transfer principle

$$(\forall^{st}\ s > 0)(\exists d > 0)(\forall \varepsilon \leqq d)(|\omega(\varepsilon)-1/\varepsilon| < s/\sqrt{\varepsilon}) \Rightarrow$$

$$(***) \qquad (\forall s > 0)(\exists d > 0)(\forall \varepsilon \leqq d)(|\omega(\varepsilon)-1/\varepsilon| < s/\sqrt{\varepsilon})$$

We might restate this entirely classical statement in the following concise way

$$\lim_{\varepsilon \downarrow 0} \sqrt{\varepsilon}|\omega(\varepsilon)-1/\varepsilon| = 0$$

This way of referring to the Transfer Principle to translate a nonstandard result into a classical result is characteristic. In fact, it is always possible to reduce nonstandard statements automatically into classical statements by Nelson's Reduction Algorithm. See [5], [6] or [4] . Notice that the above translation is far less elementary than the proofs of Propositions 1.1-1.3. Those proofs do not use very much more than just "ε is infinitesimal" (the use of the lemma of dominated approximation is a matter of convenience, and can in this concrete case be replaced by direct reasonings).

3. What is the numerical worth of nonstandard results? In view of Nelson's Reduction Algorithm the question reduces to: What is the numerical worth of classical results? The answer to this question is highly problematic, for classical mathematics is nonconstructive. Take only for example our function $f(x) = \int_0^\infty \frac{e^{-t}}{1+xt}dt$. The definition of f contains three symbols of nonconstructive entities: e, ∞ and $\int$. Also x might be nonconstructive, if, say, $x = \pi$ (what about $x = \sqrt{2}$?). It seems to me that classical results, and a fortiori nonstandard results must be subdued to numerical experimentation. But it is even not very clear to me what should be called a succesful experiment.

4. Historically, the divergent series $\sum_{n=0} (-1)^n n! \varepsilon^n$ was already studied by Euler. It

stands out as a key-example in asymptotics (see 6 below). A version of proposition 1 was proved by Stieltjes in 1886 [61]. He studied the remainders $R_{1/\varepsilon+p}(\varepsilon)$ where ε is a "quantité infiniment petite" and p is a "valeur limitée". For these indices he proved the result of proposition 1.1, and even gave a further expansion in terms of ε and p. For such expansions, see notably Dingle [43].

In the context of formal series the proposition 1.1 was stated by among others Airey [38] and Dingle [43] but it was not rigorously proved. On the other hand, the proposition follows immediately from a formula derived by L. Berg [39], who showed (in our notation)

$$R_{n-2}(\varepsilon) = T_{n-1}(\varepsilon)\left(\frac{1}{1+n\varepsilon} + \frac{1}{n\varepsilon}\cdot\frac{\theta\varepsilon}{1+2/n+1/(\varepsilon n)^2}\right) \qquad (0 < \theta < 1)$$

Hence, for $\varepsilon \simeq 0$ one has $R_{n-2}(\varepsilon) = (1+\emptyset)\frac{T_{n-1}(\varepsilon)}{1+n\varepsilon}$.

However, it seems to me that a qualitative description of the approximation of $f(\varepsilon)$ by the partial sums $S_n(\varepsilon)$, as given in§4 above is lacking in the usual presentation of the subject.

5. The propositions are easily generalized to the integrals

$$f_r(\varepsilon) = {}_0\int^\infty \frac{e^{-t}}{(1+\varepsilon t)^r}dt \qquad \text{st } r, \quad r \notin \mathbb{Z}^-$$

Repeated integration by parts yields an expansion with general term
$T_n(\varepsilon) = (-1)^n\ r(r+1)\ \ldots\ (r+n-1)\varepsilon^n$ and remainder
$R_{n-1}(\varepsilon) = (-1)^n\ r(r+1)\ \ldots\ (r+n-1)\varepsilon^n\ {}_0\int^\infty \frac{e^{-t}}{(1+\varepsilon t)^{n+r}}dt$. The formula of proposition 1.1 now becomes

$$R_{n-1}(\varepsilon) = (1+\emptyset)\ \frac{T_n(\varepsilon)}{1+(n+r)\varepsilon}$$

Its proof is almost unchanged.

6. From a theoretical point of view, the expansion $f(x) \sim {}_{k=0}\sum(-1)^k k!x^k$ is known as an <u>asymptotic expansion</u>. Very briefly we state here the classical definition of this notion, which stems from Poincaré.

An <u>asymptotic sequence</u> (in 0) is a sequence of functions $(\psi_n)_{n\in\mathbb{N}}$ such that $\lim_{x\to 0} \psi_{n+1}(x)/\psi_n(x) = 0$. The sequence of monomials $(x^n)_{n\in\mathbb{N}}$ is an example of such an asymptotic sequence. Here we stick to this example. An <u>asymptotic expansion</u> of a function $\varphi(x)$ in 0 (with respect to the asymptotic sequence $(x^n)_{n\in\mathbb{N}}$) is a formal series ${}_{k=0}\sum c_k x^k$, where the sequence of coefficients is defined by induction in the following manner

<u>Figure 6</u>:
If $\varepsilon \simeq 0$, $\varepsilon > 0$, the number $f(\varepsilon) = \int_0^\infty \frac{e^{-t}}{1+\varepsilon t}dt$ possesses a shadow expansion $\sum_{n=0} c_n \varepsilon^n$. For standard n the coefficient c_n is the standard number with minimal distance to $(-1)^n n! \int_0^\infty \frac{e^{-t}}{(1+\varepsilon t)^{n+1}}dt$, i.e. $c_n = (-1)^n n!$
Otherwise said, the coefficient c_n is the shadow of the number $(-1)^n n! \int_0^\infty \frac{e^{-t}}{(1+\varepsilon t)^{n+1}}dt$. Compare this operation of "taking the shadow" with what happened in the table: we calculated $(-1)^n n! \int_0^\infty \frac{e^{-t}}{(1+\varepsilon t)^{n+1}}dt$ for $\varepsilon = 0.0001$ and for $n = 0$ to 5, and rounded it off to the nearest integer.

$(-1)^n n! \int_0^\infty \frac{e^{-t}}{(1+\varepsilon t)^{n+1}}dt$

n	8 decimals	0 decimals
0	0.99990002	1
1	-0.99980006	-1
2	1.99940024	2
3	-5.99760120	-6
4	23.98800720	24
5	-119.92805036	-120

$$c_0 = \lim_{x \to 0} \varphi(x)$$

$$c_n = \lim_{x \to 0} \frac{\varphi(x) - \sum_{k=0}^{n-1} c_k x^k}{x^n} \qquad (n \geq 1)$$

So at every stage the remainder is compared with the next term out of the asymptotic sequence. Of course the process breaks down at the moment that some limit fails to exist. Every C^∞ function possesses an asymptotic expansion, namely its Taylor series. Asymptotic expansions may be both convergent or divergent. See also Fig. 3.

Now let us introduce through an example the nonstandard counterpart of that concept. The precise definition will be stated in the last chapter. A basic feature is that <u>numbers</u> in stead of <u>functions</u> are considered.

The nonstandard counterpart of an asymptotic sequence of functions is a sequence of numbers $(u_n)_{n\in \mathbb{N}}$ such that $u_{n+1}/u_n \simeq 0$ for all $n \in \mathbb{N}$. This is called an <u>order scale</u>. Let $\varepsilon \simeq 0$, $\varepsilon > 0$. Then $(\varepsilon^n)_{n\in\mathbb{N}}$ is an order scale; here we only consider this example. The <u>shadow</u> of a limited real number y is the (unique) standard real number x whose distance to y is infinitesimal. We write $x = {}^o y$. We now determine the <u>shadow expansion</u> of the number $f(\varepsilon)$ (with respect to the order scale $(\varepsilon^n)_{n\in\mathbb{N}}$).

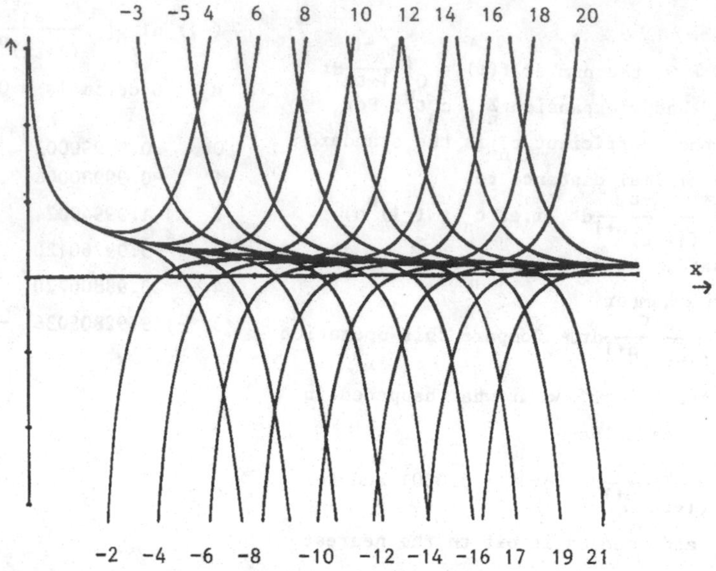

Figure 7: Regular behaviour of partial sums of asymptotic expansions.

Let us consider the function $g(x) = e^x E_1(x) = \frac{1}{x} f(\frac{1}{x})$ for $x > 0$. As is easily seen, the function has an everywhere convergent asymptotic expansion in 0:

$$g(x) = -\log x - \Gamma - \sum_{k=1}^{\infty} (-1)^k \frac{x^k}{k!k}$$

(Γ is Euler's constant, $\Gamma = 0.5772156649\ldots$). Let us write $P_n(x) = -\log x - \Gamma - \sum_{k=1}^{n} (-1)^k \frac{x^k}{k!k}$. The picture above shows the P_n from n=4 to n=21. It follows immediately from the text that the function g(x) has the everywhere divergent asymptotic expansion at infinity

$$g(x) \sim \sum_{k=0} (-1)^k \frac{k!}{x^{k+1}}$$

Let us write $Q_{-n}(x) = \sum_{k=0}^{n-1} (-1)^k \frac{k!}{x^{k+1}}$. The picture above shows the Q_{-n} for n=2 to n=17.

Notice that the graphs of the P_n "progress" to the right somewhat more slowly then the graphs of the Q_{-n} "recede" to the right. The configuration can be described with the methods of Section 4 of this chapter, or of Section 4 of Chapter III.

This is to be a formal series $\sum_{k=0} c_k \varepsilon^k$, where $(c_k)_{k\in\mathbb{N}}$ is a standard sequence (see def. 6.6 and 6.7). By definition $c_0 = {}^o(f(\varepsilon))$. We have

$$f(\varepsilon) = {}_0\int^{\infty} \frac{e^{-t}}{1+\varepsilon t} dt \simeq {}_0\int^{\infty} e^{-t} dt = 1$$

(justification of the near equality follows from the lemma of dominated approximation). The second coefficient c_1 is obtained by taking the shadow of the remainer $f(\varepsilon)-1$, compared with ε:

$$c_1 \equiv {}^o\left(\frac{f(\varepsilon)-1}{\varepsilon}\right) = {}^o\left(- {}_0\int^{\infty} \frac{e^{-t}}{(1+\varepsilon t)^2} dt\right) = -1$$

The coeffient c_2 is now the shadow of the remainder $f(\varepsilon)-(1-\varepsilon)$ compared with ε^2; we find $c_2=2$. We can take the shadow of the remainder $R_{n-1}(\varepsilon)$ compared with ε^n as long as n is standard; we find

$$c_n = {}^o\left((-1)^n n! \, {}_0\int^{\infty} \frac{e^{-t}}{(1+\varepsilon t)^{n+1}} dt\right) = (-1)^n n!$$

(justification as usual by the lemma of dominated approximation). We cannot take the shadow of $R_{n-1}(\varepsilon)/\varepsilon^n$ for unlimited n, for $R_{n-1}(\varepsilon)/\varepsilon^n$ is unlimited. However, a standard sequence of coefficients $(c_n)_{n\in\mathbb{N}}$ is obtained by simply extending the sequence $((-1)^n n!)_{st\, n\in\mathbb{N}}$ to the standard sequence $((-1)^n n!)_{n\in\mathbb{N}}$. By definition, $\sum_{k=0} (-1)^k k! \varepsilon^k$ is the shadow expansion of $f(\varepsilon)$.

§6. Comments on the method.

1. The reason why it was so easy to describe the behaviour of the remainders, and with such a precision, is in my opinion that ε was considered fixed and non-standard. Because ε was fixed our only variable was the index n; because ε was non-standard we had at our disposal ways of mathematical expression which are richer than the classical language, namely "external" notions like "limited" or "infinitesimal". The external notions are precise and well defined, but generate concepts and relations of great flexibility, notably convex proper subsets of $\mathbb{R}$ without least upper bound nor greatest lower bound. For example, the classical approach of the subject consists either in defining a functional relation between n and ε ($n=n(\varepsilon)$, so one does not dispose of information on $R_n(\varepsilon)$ for all $n \in \mathbb{N}$), or in defining no relation at all (at the cost of presenting a purely formal proof). The external setting enabled us to stay in between: we defined nonfunctional external relations between n and ε ($n\varepsilon$ limited, $n\varepsilon$ unlimited). This made it possible to present a rigorous proof, which was valid for all $n \in \mathbb{N}$, by dividing it into subcases according to the order of magnitude of n.

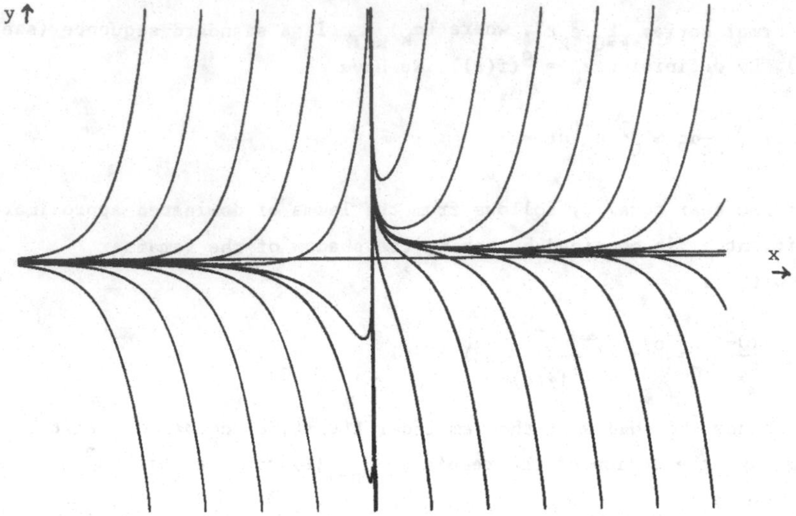

Figure 8: The function g(x) as a "stream". The function $g(x) = \frac{1}{x}f(\frac{1}{x})$ (see fig. 7) is solution of the differential equation

$$y' = y - \frac{1}{x}$$

For $x > 0$ the general solution of this equation is $y(x) = g(x)+Ce^x$ with $C \in \mathbb{R}$. The picture shows that the graph of the function g(x) acts as a (unstable) stream, organizing all other solutions: they follow the stream some time, before flowing off to positive or negative infinity. For $x < 0$ the picture shows stable streams. The configuration is typical for rational differential equations

$$y' = F(x,y) \qquad F \text{ a rational function}$$

Indeed, under fairly general conditions, its solutions are organized around a finite number of streams, which may be stable or unstable, and which have divergent asymptotic expansions at infinity, resembling that of the function g(x). See the work of M. Diener and Reeb [22] [26], Ramis and Martinet [52] [58] and F. Diener [19] [20].

2. In the proofs we used as an important tool the lemma of dominated approximation. This lemma is related to Lebesgue's dominated convergence theorem. The force of the lemma lays in the fact that an approximation of an integral is obtained using only local approximation of the integrand. For instance, in order to approximate $\int_0^\infty \frac{e^{-t}}{(1+\varepsilon t)^{n+1}}dt$ it sufficed to approximate $\frac{e^{-t}}{(1+\varepsilon t)^{n+1}}$ for limited t (which was very easy; however, for larger t much harder work must be done in order to obtain a useful approximation, the approximation $\frac{e^{-t}}{(1+\varepsilon t)^{n+1}} \simeq 0$ is too weak to conclude even convergence of the integral). The lemma of dominated approximation ensures that any, however coarse , standard integrable majoration (here e^{-t}) justifies the approximation of the integral resulting from the local approximation of the integrand.

The external notions and the lemma of dominated approximation have been the most outstanding tools in my investigations in asymptotic analysis.

2. Stirling's formula.

§1. Introduction.

Let ω be positive unlimited. The asymptotic approximation

$$\Gamma(\omega+1) = (1+\emptyset)\omega^\omega e^{-\omega}\sqrt{\omega}\sqrt{2\pi}$$

is known as Stirling's formula. Stirling's formula is a particular case of the theorem below ($\psi \equiv 1$). The theorem is an instance of what is classically known as "the Laplace method", see for instance [51]. A nonstandard version of this "method" will be proved in a general setting in chapter V (lemma 5.7). The special case below plays a key role in the estimation of the remainders of asymptotic expansions based on Laplace or Borel transforms (Chapters II and III). So it is not only for didactical reasons that we present here an individual proof.

The proof uses some basic external notions:

1. Let $\varepsilon \in \mathbb{R}$, $\varepsilon > 0$. The ε-galaxy (ε-gal) is defined by

$$\varepsilon\text{-gal} = \{u\varepsilon \mid u \in \mathbb{R} \text{ limited}\}$$

The ε-halo (ε-hal) is defined by

$$\varepsilon\text{-hal} = \{\alpha\varepsilon \mid \alpha \in \mathbb{R} \text{ infinitesimal}\}$$

The 1-galaxy is called the principal galaxy, the 1-halo is called the halo of 0.

2. Let $\varepsilon \simeq 0$, $\varepsilon > 0$. The ε-microhalo (ε-M) is defined by

$$\varepsilon\text{-M} = \{\mu \in \mathbb{R} \mid \mu < \varepsilon^n \text{ for all st } n \in \mathbb{N}\}$$

Notice that the numbers $e^{-\alpha/\varepsilon}$, $\alpha \not\gtrsim 0$, are nonzero elements of ε-M. The above external sets are part of a wider class of external sets, namely the "galaxies" and the "halos". The external sets are treated in detail in chapter IV.
3. A real function f is called <u>S-continuous</u> at a point x if $f(y) \simeq f(x)$ for all $y \simeq x$. Every standard continuous real function is S-continuous at every standard point. For more details, see any introduction to nonstandard analysis.
4. A complex function f is called of <u>S-exponential order</u> in some direction of the complex plane if there are standard constants K and C such that $|f(z)| \leq Ke^{C|z|}$ for all z in this direction.
See definition 2.8.

§2. <u>Theorem 1.4</u>: Let $\varepsilon \in \mathbb{C}$, $\varepsilon \simeq 0$. Let ψ be a Riemann integrable internal function, which is S-continuous and appreciable for limited t, and of S-exponential order in the direction of arg ε. Let ω be unlimited and such that $\omega\varepsilon$ is limited. Then

$$(1) \qquad \int_0^\infty e^{-t}t^\omega\psi(\varepsilon t)dt = (1+\emptyset)\Gamma(\omega+1)\psi(\varepsilon\omega)$$

<u>Proof</u>: The integrand takes its highest values around $t=\omega$. With the substitution $t = \omega(1+u)$ we place the peak in 0 and rescale:

$$\int_0^\infty e^{-t}t^\omega\psi(\varepsilon t)dt = \omega^\omega e^{-\omega}\omega\int_{-1}^\infty e^{-\omega(u-\log(1+u))}\psi(\varepsilon\omega+\varepsilon\omega u)du$$

After the rescaling almost all contribution to the integral is contained in the halo of 0. A new tool, namely the concentration lemma (lemma 5.6) permits us to express this fact in a precise manner:

$$\int_{-1}^\infty e^{-\omega(u-\log(1+u))}\psi(\varepsilon\omega+\varepsilon\omega u)du = \int_{-\delta}^\delta e^{-\omega(u-\log(1+u))}\psi(\varepsilon\omega+\varepsilon\omega u)du+\mu$$

where $\delta \simeq 0$ and $\mu \in 1/\omega$-M. Let us write $K(u) = e^{-\omega(u-\log(1+u))}\psi(\varepsilon\omega+\varepsilon\omega u)$. The above operations enable us to replace the integrand by an approximation. Indeed, we have on the whole interval of integration $[-\delta,\delta]$

$$K(u) = e^{-\omega\frac{u^2}{2}(1+\emptyset)}\psi(\varepsilon\omega)(1+\emptyset)$$

We approximate $\int_{-\delta}^\delta K(u)du$ with the help of the lemma of dominated approximation. In order to get a significative approximation we again have to rescale. The above approximation shows that $\{u \mid K(u) \not\simeq 0\} = 1/\sqrt{\omega}$-gal. So by the substitution $v = \sqrt{\omega}u$ we obtain an integrand which is appreciable on the whole principal galaxy:

$$\int_{-\delta}^\delta K(u)du = \frac{\psi(\varepsilon\omega)}{\sqrt{\omega}}\int_{-\delta\sqrt{\omega}}^{\delta\sqrt{\omega}}(1+\emptyset)e^{-(1+\emptyset)\frac{v^2}{2}}dv$$

Now $(1+\emptyset)e^{-(1+\emptyset)\frac{v^2}{2}} \simeq e^{-\frac{v^2}{2}}$ for all limited v and $(1+\emptyset)e^{-(1+\emptyset)\frac{v^2}{2}} \leqq 2e^{-\frac{v^2}{4}}$ for all v on the interval $[-\delta\sqrt{\omega},\delta\sqrt{\omega}]$ (which is of unlimited lenght). Hence by the lemma of dominated approximation

$$\int_{-\delta\sqrt{\omega}}^{\delta\sqrt{\omega}} (1+\emptyset)e^{-(1+\emptyset)\frac{v^2}{2}}dv \simeq \int_{-\infty}^{\infty} e^{-\frac{v^2}{2}}dv = \sqrt{2\pi}$$

For $\psi \equiv 1$ we thus obtain Stirling's formula: $\Gamma(\omega+1) = \int_0^\infty e^{-t}t^\omega dt = \omega^\omega e^{-\omega}\sqrt{\omega}\sqrt{2\pi}$. This implies (1).

N.B. Of course, if ψ is constant, or $\varepsilon = 0$, there is no upper limit to the value of ω.

§3. Comments on the proof.

1. The approximation of the integral ${}_0\int^\infty e^{-t}t^\omega\psi(\varepsilon t)dt$ is a two-parameter problem in ω and ε. But the problem would not be essentially simpler if ψ was a constant. Indeed, in the domain where the main contribution to the integral is located, namely the $\sqrt{\omega}$-galaxy around ω, the function ψ is almost constant ($\psi(\varepsilon(\omega+\sqrt{\omega}v)) \simeq \psi(\varepsilon\omega)$ for limited v). In a sense, the integral behaves like a distribution, with $e^{-t}t^\omega$ as a kernel and ψ as a test-function. See fig. 9.

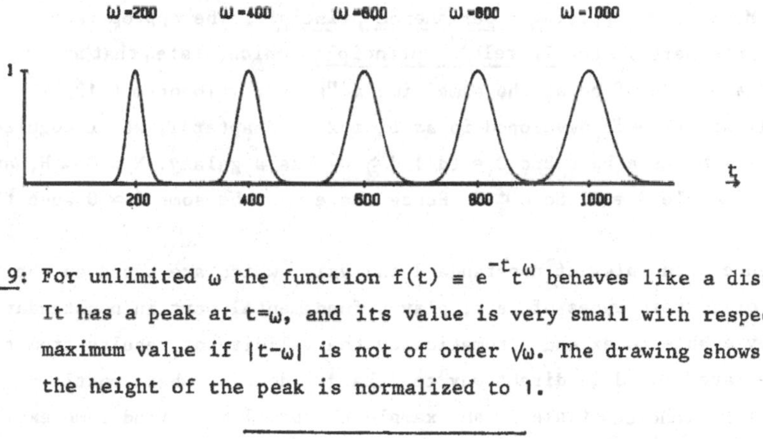

Figure 9: For unlimited ω the function $f(t) \equiv e^{-t}t^\omega$ behaves like a distribution. It has a peak at $t=\omega$, and its value is very small with respect to its maximum value if $|t-\omega|$ is not of order $\sqrt{\omega}$. The drawing shows $\frac{f(t)}{e^{-\omega}\omega^\omega}$, i.e. the height of the peak is normalized to 1.

---

2. The concentration lemma (lemma 5.6) says that the contribution to all integrals of type ${}_0\int^\infty e^{-\omega\varphi(u)}\psi(u)du$, where φ is a standard increasing function and ψ does not vary too much and grow too fast, is very highly concentrated in the halo of zero. More precisely, there exists a (somewhat large) infinitesimal δ, such that the error made by shrinking the interval of integration to $[0,\delta]$ is smaller than any standard power of $1/\omega$.

The basic interest of the concentration lemma is that now local approximation of the integrand is valid on the whole of the interval of integration . This is well illustrated by the above proof: we replaced $u-\log(1+u)$ by the first nonzero term of its Taylor series and $\psi(\varepsilon\omega+\varepsilon\omega u)$ by its approximate value $\psi(\varepsilon\omega)$. In both cases the resulting error is only infinitesimal for all $u \simeq 0$, hence certainly for all $u \in [-\delta,\delta]$. See Fig. 10. In the above proof we did not make full use of the particularity that the error μ resulting from the shortening of the interval of integration can be supposed smaller than all standard powers of $1/\omega$ ($\mu = \emptyset/\sqrt{\omega}$ would do). However, this property is useful in case a sharper approximation of the integral is desired, say a further expansion in powers of $1/\omega$.

3. Though the concentration lemma is not difficult to prove in the general case, its proof in the simple case $\int_0^\infty e^{-\omega(u-\log(1+u))}du$ is instructive. Let us define

$$\varphi(u) = u-\log(1+u)$$

$$J(d) = \int_d^\infty e^{-\omega(u-\log(1+u))}du$$

It is easy to see that $J(d) \in 1/\omega\text{-}M$ for all $d \gneqq 0$. Indeed, notice that $\varphi'(u)$ is increasing on $[d,\infty]$ and that $\varphi(d),\varphi'(d) \gneqq 0$. Hence

$$J(d) \leqq e^{-\omega\varphi(d)} \int_d^\infty e^{-\omega\varphi'(d)(u-d)} = \frac{e^{-\omega\varphi(d)}}{\omega\varphi'(d)} \in 1/\omega\text{-}M.$$

Surprisingly, we are now able to conclude immediately that there exists $\delta \simeq 0$ such that $J(\delta) \in 1/\omega\text{-}M$, i.e. by applying a <u>permanence principle</u>. The appropriate permanence principle here is the <u>Fehrele[(1)] principle</u>, which states that an external set which is a "halo" cannot be at the same time a "galaxy" (Theorem 4.12). Indeed, on behalf of criteria developed in section 2 of Chapter IV, we recognize $H = \{d \mid J(d) \in 1/\omega\text{-}M\}$ as a halo and $G = \{d \mid d \gneqq 0\}$ as a galaxy. Now $G \subset H$, and by the Fehrele principle $G \neq H$. So $G \subsetneqq H$. Hence there must be some $\delta \simeq 0$ such that $J(\delta) \in 1/\omega\text{-}M$.

The permanence principles of nonstandard analysis, which are based on the incompatibility of certain types of sets, play a fundamental part in nonstandard asymptotics. They enable to extend automatically the validity of results from the domain where they are proved in direct way to a larger domain, where mostly a direct proof is more laborous (Take as an example the problem to find some explicit $\delta \simeq 0$ such that $J(\delta) \in 1/\omega\text{-}M$). The proof of the lemma of dominated approximation uses a permanence principle. Permanence principles are frequently used in the nonstandard approach to the study of singular perturbations of differential equations, notably in the problem of matching, i.e. of the global description of a solution, whose asymptotic approximation is given in different contiguous domains without sharp boundaries. See the work of Benoit, F. and M. Diener, Callot, Lutz and T. Sari.

---

(1) "A halo isch ke galaxie, das hat de Fehrele gesajd", Alsatian traditional

u	$e^{-\omega(u-\log(1+u))}$	v	$e^{-(v-\log(1+v))}$	$e^{-v^2/2}$
-0.9	1.953E-381	-0.6	0.7288	0.8353
-0.8	1.954E-220	-0.5	0.8244	0.8825
-0.7	1.602E-137	-0.4	0.8951	0.9231
-0.6	1.406E-086	-0.3	0.9449	0.9568
-0.5	3.744E-053	-0.2	0.9771	0.9802
-0.4	8.282E-031	-0.1	0.9947	0.9950
-0.3	4.135E-016	0.0	1.0000	1.0000
-0.2	5.225E-007	0.1	0.9953	0.9950
-0.1	3.507E-002	0.2	0.9825	0.9802
0.0	1.000E+000	0.3	0.9631	0.9560
0.1	5.334E-002	0.4	0.9384	0.9231
0.2	1.590E-005	0.5	0.9098	0.8825
0.3	6.087E-011	0.6	0.8781	0.8353
0.4	5.707E-018			
0.5	2.188E-026			
0.6	5.183E-036			
0.7	1.064E-046			
0.8	2.501E-058			
0.9	8.518E-071			
1.0	5.125E-084			

δ	$\sqrt{\frac{\omega}{2\pi}}\int_{-\delta}^{\delta} e^{-\omega(u-\log(1+u))}du$	$\frac{1}{\sqrt{2\pi}}\int_{-\delta\sqrt{\omega}}^{\delta\sqrt{\omega}} e^{-v^2/2}dv$
0.05	0.78861387	0.78870045
0.10	0.98737242	0.98758067
0.15	0.99990781	0.99982317
0.20	1.00013184	0.99999943
0.25	1.00013334	1.00000000
0.30	1.00013334	1.00000000

<u>Figure 10</u>:(Left). The function $e^{-\omega(u-\log(1+u))}$ becomes very small at a moderately little distance to the origine. (Right) Approximation of $e^{-(v-\log(1+v))}$ by $e^{-\frac{v^2}{2}}$ near the origin. (Below) Approximation of $\sqrt{\frac{\omega}{2\pi}}\int_{-\delta}^{\delta} e^{-\omega(u-\log(1+u))}$ du by $\frac{1}{\sqrt{2\pi}}\int_{-\delta\sqrt{\omega}}^{\delta\sqrt{\omega}} e^{-\frac{v^2}{2}}dv$. Both integrals are close to 1 (respectively the approximate value of $\sqrt{\frac{\omega}{2\pi}}\int_{-1}^{\infty} e^{-\omega(u-\log(1+u))}du$ and the exact value of $\frac{1}{\sqrt{2\pi}}\int_{-\infty}^{\infty} e^{-\frac{v^2}{2}}dv$) already for small δ. The number ω has throughout the value $\omega = 100$.

4. Besides the concentration lemma another nonstandard tool is used in the proof of Theorem 1.4. It consists of a method to get a substitution which turns a limited integrand (here K(u)) into an integrand who takes limited values on a set which is contained in the principal galaxy, but which is larger than the halo of zero (here $(1+\emptyset)e^{-(1+\emptyset)\frac{v^2}{2}}$). Only then the lemma of dominated approximation gives a significative approximation. Notice that the approximation

$$K(u) \simeq \begin{cases} \psi(\varepsilon\omega) & u = 0 \\ 0 & \text{otherwise} \end{cases}$$

is correct, but the approximation $\int_{-1}^{\infty} K(u)du \simeq 0$ is too weak. The method is based on the fact that there exists a (linear) mapping f from the principal galaxy onto the galaxy $A = \{u \mid K(u) \not\simeq 0\}$. In order to find such a mapping explicitly we must determine the shape of the galaxy $A = \{u \mid K(u) \not\simeq 0\}$. The problem to find the shape of an (convex) external set is called an external equation (More about external equations in Section 4 of this chapter; they are treated in detail in Section 5 of Chapter IV). Here the solution of the external equation is simple.
Indeed, we saw by inspection that $A = 1/\sqrt{\omega}$-gal: so we can choose $f(u) = u/\sqrt{\omega}$. Then the substitution $v = \sqrt{\omega}u$ transforms K(u) into an integrand with the required properties. This method of recognition of the set where the integrand takes appreciable values, which is somewhat elaborate in simple cases, guided me in more complex situations, notably in the proof of proposition 3.10.

3. The expansion at infinity of the Besselfunctions.

We study the approximation of the Besselfunctions $J_p(z)$ by the partial sums of their classical divergent expansion at infinity. Though the mathematical expressions in question will be somewhat involved, the nonstandard asymptotic analysis of the remainders presents no major difficulties.

§1. Review and notation.

We first present one of the classical methods to obtain the terms of the expansion at infinity of the Besselfunctions $J_p(z)$. They will be derived from the expansion in 1/z of the Hankelfunctions $H_p^{(1)}(z)$ and $H_p^{(2)}(z)$, which in turn are derived from the Taylor expansions in 0 of the functions $(1+t)^{p-\frac{1}{2}}$ respectively $(1-t)^{p-\frac{1}{2}}$. The remainders of the expansions at infinity of the functions $J_p(z)$ will be obtained along the same lines.

The Hankelfunctions of order $p > -\frac{1}{2}$ are defined by

$$H_p^{(1)}(z) = \sqrt{\frac{2}{\pi z}}\,\frac{e^{i(z-\frac{\pi}{2}(p+\frac{1}{2}))}}{\Gamma(p+\frac{1}{2})}\int_0^\infty e^{-t}t^{p-\frac{1}{2}}\left(1+\frac{it}{2z}\right)^{p-\frac{1}{2}}dt \qquad -\frac{\pi}{2} < \arg z < \frac{3\pi}{2}$$

$$H_p^{(2)}(z) = \sqrt{\frac{2}{\pi z}}\,\frac{e^{-i(z-\frac{\pi}{2}(p+\frac{1}{2}))}}{\Gamma(p+\frac{1}{2})}\int_0^\infty e^{-t}t^{p-\frac{1}{2}}\left(1-\frac{it}{2z}\right)^{p-\frac{1}{2}}dt \qquad -\frac{3}{2}\pi < \arg z < \frac{\pi}{2}$$

The relation between the Hankelfunctions and the Besselfunctions is given by

$$J_p(z) = \frac{H_p^{(1)}(z)+H_p^{(2)}(z)}{2}$$

We write

$$\Theta = z-\frac{\pi}{2}(p+\tfrac{1}{2})$$

$$(r)_k = r(r+1)\ \ldots\ (r+k-1) \qquad \text{("Pochhammer's symbol")}$$

Let $t_k(t)$ be the term of degree k of the Taylor expansion of $(1+t)^{p-\frac{1}{2}}$ in 0. Then

$$t_k(t) = \frac{(-1)^k(\frac{1}{2}-p)_k}{k!}t^k$$

The k-th term $T_k^{(1)}$ of the development in $1/z$ of $H_p^{(1)}(z)$ is defined by

$$\begin{aligned} T_k^{(1)}(z) &= \sqrt{\frac{2}{\pi z}}\,\frac{e^{i\Theta}}{\Gamma(p+\frac{1}{2})}\int_0^\infty e^{-t}t^{p-\frac{1}{2}}t_k\left(\frac{it}{2z}\right)dt \\ &= \sqrt{\frac{2}{\pi z}}\,\frac{e^{i(\Theta-\frac{k}{2}\pi)}(\frac{1}{2}-p)_k}{(2z)^k k!\,\Gamma(p+\frac{1}{2})}\int_0^\infty e^{-t}t^{k+p-\frac{1}{2}}dt. \\ &= \sqrt{\frac{2}{\pi z}}\,\frac{e^{i(\Theta-\frac{k}{2}\pi)}}{(2z)^k}\,\frac{(\frac{1}{2}-p)_k}{k!}\,\frac{\Gamma(k+p+\frac{1}{2})}{\Gamma(p+\frac{1}{2})} \\ & A_k e^{i(\Theta-\frac{k}{2}\pi)} \end{aligned}$$

with

$$A_k = \sqrt{\frac{2}{\pi z}}\,\frac{(\frac{1}{4}-p^2)(\frac{9}{4}-p^2)\cdot\ \ldots\ \cdot((k-\frac{1}{2})^2-p^2)}{k!(2z)^k}$$

The corresponding terms of the expansion of $H_p^{(2)}(z)$ are derived analogously from the Taylor expansion of the function $(1-t)^{p-\frac{1}{2}}$. One has

$$T_k^{(2)}(z) = A_k e^{-i(\Theta-\frac{k}{2}\pi)}$$

The k-th term of the expansion in $1/z$ of the Besselfunction J_p is then defined by

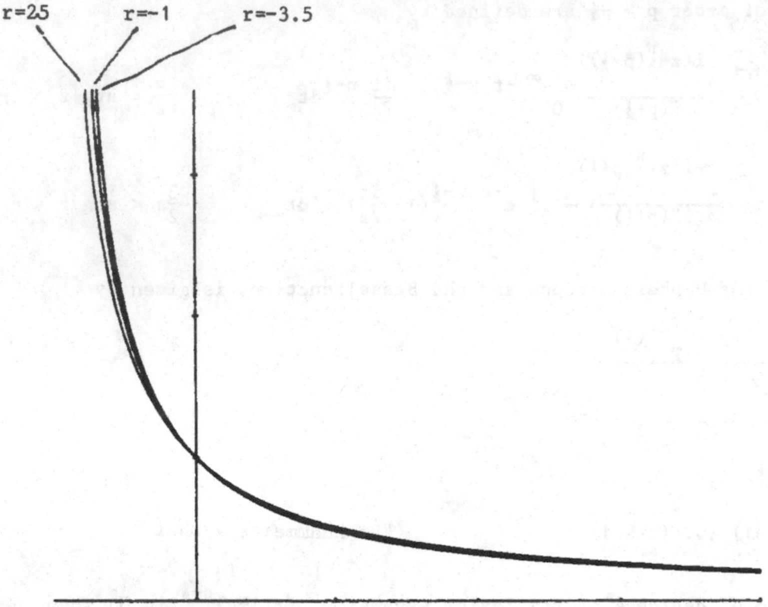

Figure 11: The quotients $R_{n-1}(x)/T_n(x)$ associated to the Taylorseries of the functions $(1+x)^r$, for r=2.5, r=-1, respectively r=-3.5. The index n has the value 50. Theoretically, the quotients are almost equal for limited x and unlimited r.

$$T_k(z) = \frac{T_k^{(1)}(z)+T_k^{(2)}(z)}{2} = A_k \cos(\Theta-\frac{k}{2}\pi)$$

Let $r_{n-1}(t) = (1+t)^{p-\frac{1}{2}}-\sum_{k=0}^{n-1} t_k(t)$ be the remainder after truncating the Taylor-expansion of $(1+t)^{p-\frac{1}{2}}$ at degree n-1. Then the remainder $R_{n-1}^{(1)}(z)$ of the Hankelfunction $H_p^{(1)}(z)$ is given by

$$R_{n-1}^{(1)}(z) = \sqrt{\frac{2}{\pi z}}\frac{e^{i\Theta}}{\Gamma(p+\frac{1}{2})} \int_0^\infty e^{-t}t^{p-\frac{1}{2}}r_{n-1}(\frac{it}{2z})dt \tag{2}$$

The remainder $R_{n-1}^{(2)}(z) = H_p^{(2)}(z)-\sum_{k=0}^{n-1} T_k^{(2)}(z)$ of the second Hankelfunction is obtained analogously. Then we have for the remainder $R_{n-1}(z) = J_p(z)-\sum_{k=0}^{n-1} T_k(z)$ of the Bessel-function

$$R_{n-1}(z) = \frac{R_{n-1}^{(1)}(z)+R_{n-2}^{(2)}(z)}{2}$$

§2. Results.

The results below show that in spite of the complexity of the above formula's there exists in each case a simple asymptotic relation between the remainder and the first neglected term of the expansion.

Lemma 1.5: Let $(c_n)_{n\in\mathbb{N}}$ be a standard sequence such that $\frac{c_{n+1}}{c_n} = 1+\frac{a}{n}+o(1/n)$ for $n \to \infty$, then $c_\omega = \omega^{a+\emptyset}$ for all unlimited $\omega \in \mathbb{N}$.

Proposition 1.6: Let $f(t) = (1+t)^r$. Let $t_k(t)$ be the k-th term of its Taylor-expansion and $r_{n-1}(t) = f(t)-\sum_{k=0}^{n-1} t_k(t)$ be the corresponding remainder. Then

(i) If $n \in \mathbb{N}$ is limited, then $r_{n-1}(t) = (1+\emptyset)t_n(t)$ for all $t \simeq 0$.

(ii) If $\omega \in \mathbb{N}$ is unlimited, then $r_{\omega-1}(t) = (1+\emptyset)\frac{t_\omega(t)}{1+t}$ for all t, except for $|t| \gtrsim 1$ in the directions $\arg t \simeq \pi$.

See Fig. 11.

Proposition 1.7: Let $p \gneq -\frac{1}{2}$ be limited, let $\omega \in \mathbb{N} - \{0\}$, and let $z \in \mathbb{C}$ be unlimited.

(i) If $|\arg z| \lneq \pi/2$ then $R^{(1)}_{\omega-1}(z) = (1+\emptyset)\,\frac{T^{(1)}_\omega(z)}{1+i\omega/(2z)}$

(ii) If $|\arg z| \lneq \pi/2$ then $R^{(2)}_{\omega-2}(z) = (1+\emptyset)\,\frac{T^{(2)}_\omega(z)}{1-i\omega/(2z)}$

Theorem 1.8: Let $p \gneq -\frac{1}{2}$. (i) If $z \in \mathbb{C}$ is unlimited, with $|\arg z| \lneq \pi/2$, and $\omega \geqq 1$, then

(3) $$R_{\omega-1}(z) = \frac{(1+\emptyset)T_\omega(z)+(1+\emptyset)T_{\omega+1}(z)}{1+(\omega/(2z))^2}$$

(ii) If x is real unlimited and ω/x is limited, then

(4) $$R_{\omega-1}(x) = (1+\emptyset)\frac{A_\omega}{\sqrt{1+(\omega/(2x))^2}}(\cos(\Theta-\omega\tfrac{\pi}{2}-\operatorname{arctg}\tfrac{\omega}{2x})+\emptyset)$$

§3. Proofs.

Proof of Lemma 1.5: Let $\omega \in \mathbb{N}$ be unlimited. If n is standard, then $\log n/\log\omega \simeq 0$ and $\log c_n/\log\omega \simeq 0$. Applying a permanence principle we automatically conclude that there exists nonstandard ν such that $\log\nu/\log\omega \simeq 0$ and $\log c_\nu/\log\omega \simeq 0$. The permanence principle in question is Robinson's sequential lemma (page 113, also the Fehrele principle could be used). Now

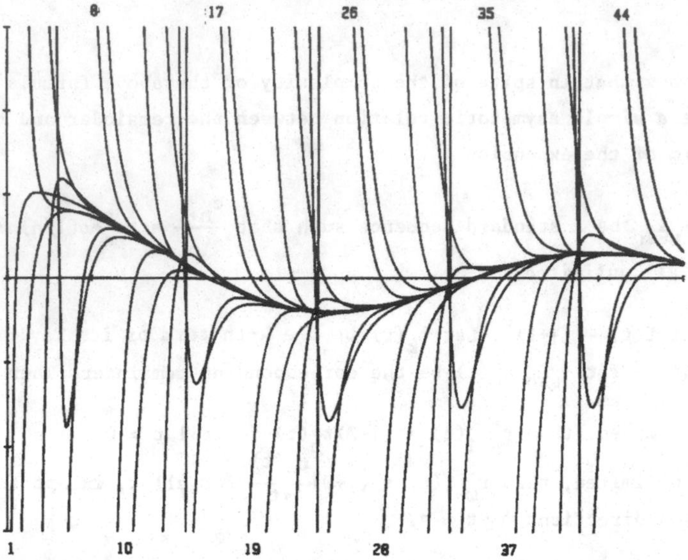

<u>Figure 12</u>: Partial sums $S_n(x) = \sum_{k=0}^{n} \sqrt{\frac{2}{\pi x}} \frac{(1/2)_k (1/2)_k}{k!(2x)^k} \cos(x - \frac{\pi}{4} - \frac{k\pi}{2})$ of the divergent expansion of the Besselfunction $J_0(x)$. The approximation is exact at points which are nearly equally spaced at a distance of about π (exercise .....)

$$c_\omega = \frac{c_\omega}{c_{\omega-1}} \cdot \ldots \cdot \frac{c_{\nu+1}}{c_\nu} \cdot c_\nu = \exp\left(\sum_{n=\nu}^{\omega-1} \log\left(1+\frac{a}{n}+\frac{\emptyset}{n}\right) + \log c_\nu\right)$$

$$= \exp\left(\sum_{n=\nu}^{\omega-1} \frac{(a+\emptyset)}{n} + \emptyset \log \omega\right) = \exp\left((1+\emptyset) \int_\nu^{\omega-1} a/x \, dx + \emptyset . \log \omega\right)$$

$$= \exp((1+\emptyset)(\log(\omega-1) - \log \nu))a + \emptyset \log \omega) = \exp((a+\emptyset) \log \omega) = \omega^{a+\emptyset}.$$

<u>Proof of proposition 1.6</u>: (i) By Lagrange's formula of the remainder of Taylor-expansions we have

$$r_{n-1}(t) = t_n(t)(1+\Theta t)^{r-n} \qquad 0 < \Theta < 1$$

So if $t \simeq 0$ and n is limited, then $(1+\Theta t)^{r-n} \simeq 1$. Hence $r_{n-1}(t) = (1+\emptyset) t_n(t)$.
(ii) Put $c_k = (-1)^k (-r)_k / k!$. Because $c_{k+1}/c_k = (r-k)/(k+1)$ we have for unlimited k the approximation $c_{k+1}/c_k = -1 + (r+1+\emptyset)/k$ and thus by lemma 1.5 the approximation $c_k = k^{-(r+1+\emptyset)}$.
We distinguish three cases: 1. $|t| \lessapprox 1$, 2. $|t| \gtrapprox 1$, 3. $|t| \simeq 1$, $t \not\simeq -1$.

1. $|t| \lessapprox 1$. It follows from the convergence of the series that

$$r_{\omega-1}(t) = t_\omega(t) \sum_{n=0}^{\infty} \frac{c_{\omega+n}}{c_\omega} t^n$$

Now $\frac{c_{\omega+n}}{c_\omega} t^n \simeq (-t)^n$ for all limited n and $\left|\frac{c_{\omega+n}}{c_\omega} t^n\right| \leqq \left(\frac{1+^{o}|t|}{2}\right)^n$ for all n. Hence, by the lemma of dominated approximation

$$\sum_{n=0}^{\infty} c_n t^n \simeq \sum_{n=0}^{\infty} (-t)^n = \frac{1}{1+t}$$

This implies that $r_{\omega-1}(t) = \frac{t_\omega(t)}{1+t}(1+\emptyset)$

2. $|t| \gtrapprox 1$. The next formula is an identity

$$r_{\omega-1}(t) = f(t) - t_{\omega-1}(t) \sum_{n=0}^{\omega-1} \frac{c_{\omega-n-1}}{c_{\omega-1}}(1/t)^n$$

Let us write $u_n = \frac{c_{\omega-n-1}}{c_{\omega-1}}(1/t)^n$. We remark that, when $\omega-n$ is limited,

$$u_n = \frac{c_{\omega-n-1} t^{\omega-n} \omega^{r+1+\emptyset}}{t^\omega} \simeq 0$$

So $\sum_{n=\omega-m}^{\omega-1} u_n \simeq 0$, whenever m is limited and by Robinson's lemma there exists unlimited ν such that $\sum_{n=\omega-\nu+1}^{\omega-1} u_n \simeq 0$. We approximate $\sum_{n=1}^{\omega-\nu} u_n$ by $\sum_{n=0}^{\infty}(-1/t)^n = \frac{1}{1+1/t}$ on behalf the lemma of dominated approximation: we have $u_n \simeq (-1/t)^n$ for limited n and $u_n \leqq \left(\frac{2}{1+|t|}\right)^n$ for all n. Combining

$$r_{\omega-1}(t) = f(t) - t_{\omega-1}(t)\frac{t}{1+t}(1+\emptyset) = f(t) + \frac{t_\omega(t)}{1+t}(1+\emptyset) = \frac{t_\omega(t)}{1+t}(1+\emptyset)$$

3. $|t| \simeq 1$, $t \not\simeq -1$. By the formula of the remainder of Taylorexpansions in integral form $\left(r_{\omega-1}(t) = \frac{t^{\omega-1}}{(\omega-1)!} \int_0^t (t-v)^{\omega-1} f^{(\omega)}(v)dv\right)$

$$r_{\omega-1}(t) = t_\omega(t) \int_0^\omega \left(1-\frac{u}{\omega}\right)^{\omega-1}\left(1+\frac{tu}{\omega}\right)^{r-\omega} du$$

Let us write $I(u) = \left(1-\frac{u}{\omega}\right)^{\omega-1}\left(1+\frac{tu}{\omega}\right)^{r-\omega}$. If u is limited, then $I(u) \simeq e^{-(1+t)u}$. Put $x = \mathrm{re}\, t$. Notice that $-1 \lnapprox x \lessapprox 1$ and $|t-1| \geqq |x-1|$. Then, for all $u \leqq \omega$

$$|I(u)| = \left|1+\frac{tu}{\omega}\right|^{r-1}\left|\frac{1-u/\omega}{1+tu/\omega}\right|^{\omega-1}$$

$$\leqq K\left(1-\frac{1+x}{1+xu/\omega}\cdot\frac{u}{\omega}\right)^{\omega-1} \qquad (\text{st } K)$$

$$\leqq K\left(1-\frac{1+x}{3}\cdot\frac{u}{\omega}\right)^{\omega-1} \leqq K' e^{\frac{-1+^{o}x}{6}u} \qquad (\text{st } K')$$

Now it follows from the lemma of dominated approximation that $_0\int^{\omega} I(u)du \simeq \int_0^{\infty} e^{-(1+t)u}du = \frac{1}{1+t}$. Hence $r_{\omega-1}(t) = (1+\emptyset)\frac{t_\omega(t)}{1+t}$.

Proof of proposition 1.7: (i) We distinguish three cases: 1. ω limited, 2. ω unlimited, ω/z limited, 3. ω unlimited, ω/z unlimited.

1. ω limited. It follows from Watson's lemma (Theorem 2.9) that $R_{\omega-1}^{(1)}(z) = (1+\emptyset)T_\omega^{(1)}(z)$ for all limited ω. Then also $R_{\omega-1}^{(1)}(z) = (1+\emptyset)\frac{T_\omega^{(1)}(z)}{1+i\omega/(2z)}$

2. ω unlimited, ω/z limited. As $|\arg z| \lessapprox \pi/2$, by proposition 1.6 the formula

$$r_{\omega-1}(it/(2z)) = (1+\emptyset)\frac{t_\omega(it/(2z))}{1+it/(2z)}$$

holds for all $t \geqq 0$. Then using (2) and Theorem 1.4

$$R_{\omega-1}^{(1)}(z) = \sqrt{\frac{2}{\pi z}}\,\frac{e^{i(\theta-\frac{\omega\pi}{2})}(\frac{1}{2}-p)_\omega}{(2z)^\omega\,\omega!\,\Gamma(p+\frac{1}{2})}\;{}_0\int^\infty \frac{e^{-t}t^{\omega+p-\frac{1}{2}}}{1+it/(2z)}(1+\emptyset)dt$$

$$= (1+\emptyset)\sqrt{\frac{2}{\pi z}}\,\frac{e^{i(\theta-\frac{\omega\pi}{2})}(\frac{1}{2}-p)_\omega\Gamma(\omega+p+\frac{1}{2})}{(2z)^\omega\omega!\Gamma(p+\frac{1}{2})(1+i(\omega+p-\frac{1}{2})/(2z))}$$

$$= (1+\emptyset)\,\frac{T_\omega^{(1)}(z)}{1+i\omega/(2z)}$$

3. ω unlimited, ω/z unlimited. The proof is similar to case (ii) of the proof of proposition 1.1.

(ii) The proof is analogous to the proof of (i).

Proof of Theorem 1.8: (i) We distinguish two cases: 1. $\omega/z \simeq 0$, 2. $|\omega/z| \gtrapprox 0$.

1. $\omega/z \simeq 0$. Though $R_{\omega-1}^{(1)}(z) = (1+\emptyset)T_\omega^{(1)}(z)$ and $R_{\omega-1}^{(2)}(z) = (1+\emptyset)T_\omega^{(2)}(z)$ we may not write $R_{\omega-1}(z) = (1+\emptyset)T_\omega(z)$. This may be false notably if $\cos(\theta-\frac{\omega\pi}{2}) = 0$: then $T_\omega(z) = 0$, while not necessarily $R_{\omega-1}(z) = 0$. Instead of expressing the remainder as a function of the first neglected term we will express it as a function of the first and second neglected terms. Notice that

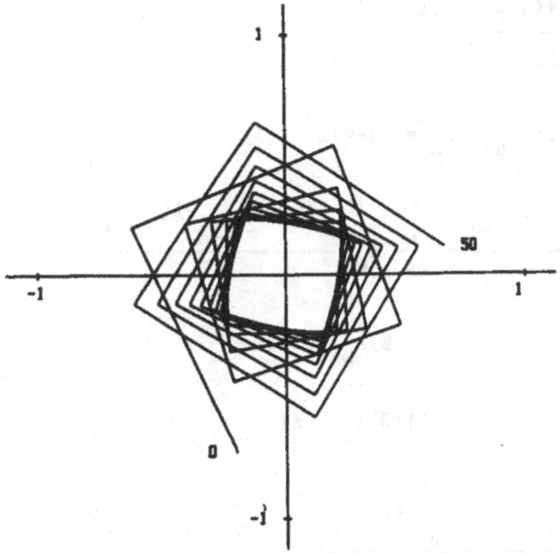

Figure 13: The Hankel function $H_0(x) = \sqrt{\frac{2}{x}}\frac{e^{-i(x-\frac{\pi}{4})}}{\pi}\ {}_0\int^{\infty}\frac{e^{-t}}{\sqrt{t(1+\frac{it}{2x})}}dt$ approached by the partial sums $S_n^{(1)}(x) = \sqrt{\frac{2}{\pi x}}e^{-i(x-\frac{\pi}{4})}\sum_{k=0}^{n}\frac{(\frac{1}{2})_k(\frac{1}{2})_k}{k!(2x)^k}e^{\frac{-ik\pi}{2}}$ of its divergent asymptotic expansion at infinity, x = 10. The picture shows for n=0 to 50 the remainders $R_n^{(1)}(x)$ -joined by straight lines-, under a radial exponential "Benoit-microcoscope". I.e., instead of $z = |z|e^{i\varphi}$ we plotted $|z|^{\varepsilon}e^{i\varphi}$ where $\varepsilon = 1/20$.

$$R_{\omega-1}^{(1)}(z) = T_\omega^{(1)}(z)+(1+\emptyset)T_{\omega+1}^{(1)}(z)$$

$$R_{\omega-1}^{(2)}(z) = T_\omega^{(2)}(z)+(1+\emptyset)T_{\omega+1}^{(2)}(z)$$

Hence

$$R_{\omega-1}(z) = \frac{R_{\omega-1}^{(1)}(z)+R_{\omega-1}^{(2)}(z)}{2}$$

$$= \frac{T_\omega^{(1)}(z)+T_\omega^{(2)}(z)}{2} + \frac{T_{\omega+1}^{(1)}(z)+T_{\omega+1}^{(2)}(z)}{2} +$$

$$+ \emptyset.\frac{T_{\omega+1}^{(1)}(z)+T_{\omega+1}^{(2)}(z)}{2} + \emptyset.T_{\omega+1}^{(2)}(z)$$

$$= (1+\emptyset)T_\omega(z)+(1+\emptyset)T_{\omega+1}(z)$$

$$\left(= \frac{(1+\emptyset)T_\omega(z)+(1+\emptyset)T_{\omega+1}(z)}{1+(\omega/(2z)^2)}\right)$$

2. $|\omega/z| \not\geq 0$. Notice that $A_{\omega+1}/A_\omega = (1+\emptyset)\omega/(2z)$. So

$$\frac{-i\omega}{2z}\,\frac{T_\omega^{(1)}(z)-T_\omega^{(2)}(z)}{2} = \frac{\omega}{2z}\circ A_\omega\,\frac{e^{i(\Theta-\frac{\omega\pi}{2})}-e^{-i(\Theta-\frac{\omega\pi}{2})}}{2i}$$

$$= (1+\emptyset)A_{\omega+1}\sin(\Theta-\frac{\omega\pi}{2})$$

$$= (1+\emptyset)T_{\omega+1}(z)$$

Hence

$$R_{\omega-1}(z) = \frac{R_{\omega-1}{}^{(1)}(z)+R_{\omega-1}{}^{(2)}(z)}{2}$$

$$= \frac{(1+\emptyset)T_\omega{}^{(1)}(z)}{2(1+i\omega/(2z))} + \frac{(1+\emptyset)T_\omega{}^{(2)}(z)}{2(1-i\omega/(2z))}$$

$$= \frac{(1+\emptyset)\left(\frac{T_\omega{}^{(1)}(z)+T_\omega{}^{(2)}(z)}{2} - \frac{i\omega}{2z}\cdot\frac{T_\omega{}^{(1)}(z)-T_\omega{}^{(2)}(z)}{2}\right)+\emptyset\circ T_\omega{}^{(2)}(z)}{1+(\omega/(2z))^2}$$

$$= \frac{(1+\emptyset)T_\omega(z)+(1+\emptyset)T_{\omega+1}(z)+\emptyset\circ T_\omega{}^{(2)}(z)}{1+(\omega/(2z))^2}$$

$$= \frac{(1+\emptyset)T_\omega(z)+(1+\emptyset)T_{\omega+1}(z)}{1+(\omega/(2z))^2}$$

(ii). If $z=x$ is real, then $R_{\omega-1}{}^{(1)}(x)$ and $R_{\omega-1}{}^{(2)}(x)$ are conjugated. Then by proposition 1.7.:

$$R_{\omega-1}(z) = 2\ \mathrm{re}\ R_{\omega-1}{}^{(1)}(z) = 2\ \mathrm{re}\left|\frac{T_\omega^{(1)}(z)}{1+i\omega/2z}(1+\emptyset)\right|$$

$$= \frac{A_\omega}{\sqrt{1+(\omega/2x)^2}}(\cos(\Theta-\frac{\omega\pi}{2}-\operatorname{arctg}\frac{\omega}{2x})+\emptyset)$$

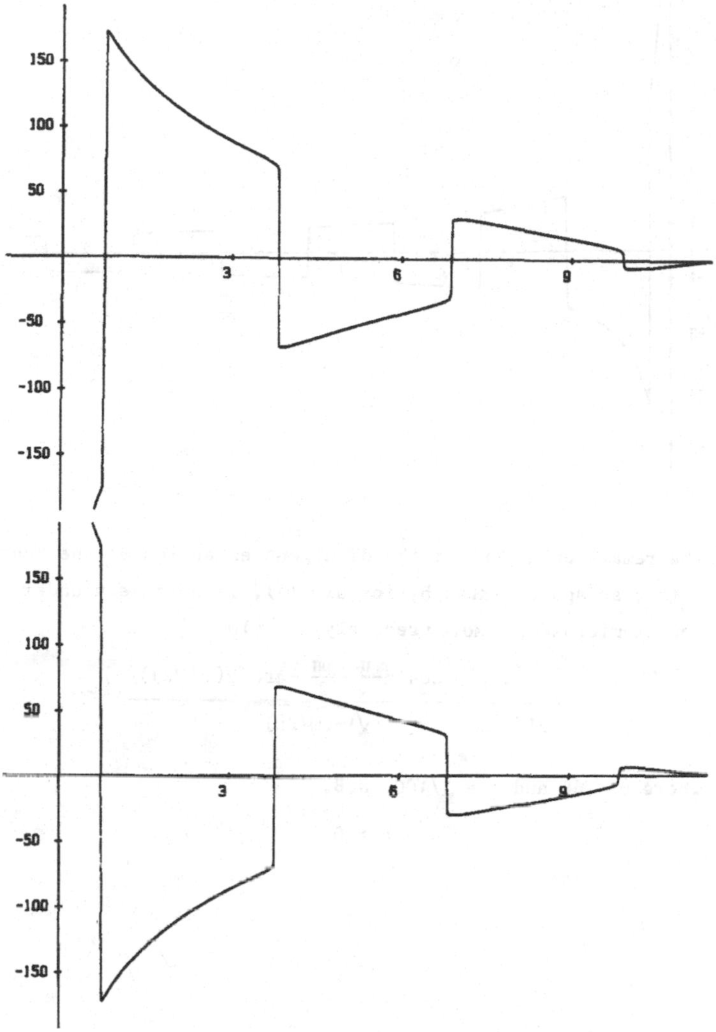

Figure 14 (above): Partial sum $S_{65}(x) = \sum_{k=0}^{65} \sqrt{\frac{2}{\pi x}} \frac{(\frac{1}{2})_k(\frac{1}{2})_k}{k!(2x)^k} \cos(x-\frac{\pi}{4}-\frac{k\pi}{2})$ of the divergent expansion of the Besselfunction $J_0(x)$, now on a logarithmical scale. More precisely, the drawing shows the graph of

$$F(x) = \begin{cases} \log S_{65}(x) + 1 & S_{65}(x) > 1 \\ S_{65}(x) & -1 \leqq S_{65}(x) \leqq 1 \\ -\log(-S_{65}(x)) - 1 & S_{65}(x) < -1 \end{cases}$$

(under): Remainder $R_{65}(x)$ associated to the same divergent expansion of the Besselfunction as approximated by formula (4), which runs in this case

$$R_{\omega-1}(x) = \frac{(1+\emptyset)(\frac{1}{2})_\omega(\frac{1}{2})_\omega}{\omega!(2x)^\omega} \frac{\cos((x-\frac{\pi}{4}-\frac{\omega\pi}{2}-\operatorname{arctg}\omega/2x)+\emptyset)}{\sqrt{1+(\omega/2x)^2}}$$

The graph is drawn with the same convention as above.

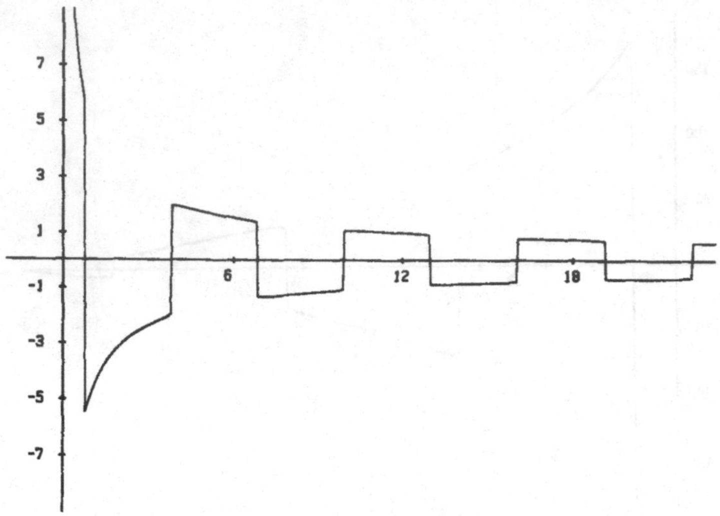

Figure 15: The remainder $R_{65}(x)$ of the divergent expansion of the Besselfunction $J_0(x)$, as approximated by formula (4), is here seen under an exponential Benoit-microscope. More precisely, we plotted

$$\left(\frac{(\frac{1}{2})^{\omega}(\frac{1}{2})^{\omega}}{\omega!/(2x)^{\omega}}\frac{\cos(x-\frac{\pi}{4}-\frac{\omega\pi}{2}-\mathrm{arctg}(\omega/2x))}{\sqrt{1+(\omega/2x)^2}}\right)^{[\varepsilon]}$$

where $\omega = 66$ and $\varepsilon = 1/100$. N.B.

$$a^{[\varepsilon]} \equiv \begin{cases} a^{\varepsilon} & a > 0 \\ -(-a)^{\varepsilon} & a < 0 \end{cases}$$

§4. Comments on the results.

1) Again we were confronted with a two parameter problem. The one-parameter problem $\omega = 2x$ is often treated in literature. See for instance Watson [62], Jeffreys [49] and Holz [48].

2) The expansions in $1/z$ of the Hankelfunctions are asymptotic expansions. It follows from the propositions 1.1 and 1.7, that the approximation properties of these expansions and that of the exponential integral are similar. In both cases a nearly optimal approximation result is attained by summing to the index of the smallest term; here this index is at limited distance from $\frac{1}{2|z|}$. In the case of the exponential

integral the partial sums oscillated around the exact value; here they turn around with angular velocity depending on arg z; for instance if z is real and positive, this angular velocity is nearly equal to $\pm\pi/2$ (see Fig. 13).

3). Due to the factor cosinus, the expansions in $1/z$ of the Besselfunctions are neither asymptotic expansions, nor shadow expansions in the case z is unlimited; they even vanish for some values of z. See Fig.12. However, they still roughly follow the evolution of the first neglected term, as shown by the formula's (3) and (4). See Fig. 12, 14 and 15.

In chapter VI we made an attempt to integrate the expansions of the Besselfunctions in a theory which is more general than that of the shadow expansions.

4. The remainder associated to the geometrical series $\frac{1}{1+t} = \sum_{n=0}^{\infty} (-1)^n t^n$ is known exactly. Indeed, $r_{n-1}(t) = \frac{t_n(t)}{1+t}$. This formula is valid both inside and outside the convergence disk. It is quite remarkable that this formula remains, asymptotically, valid in the more general case of the expansions $\frac{1}{(1+t)^r} = \sum_{n=0}^{\infty} \frac{(-1)^n (r)_n t^n}{n!}$, again both inside and outside the convergence disk. See Fig. 11.

5. The property

$$\frac{\Gamma(\omega+x)}{\Gamma(\omega)} = (1+\emptyset)\omega^x \qquad \text{(x limited)}$$

is well-known, and can be derived from Stirling's formula (see Fig. 16). In a sense, lemma 1.5 is a generalization of this property.

§ 5. Comments on the proofs.

1. In an earlier presentation [12] I used up to 23 different symbols for infinitesimals. This motivated me to introduce the symbol $\emptyset$.

2. By cutting $\mathbb{C}$ or $\mathbb{N}$ into external slices we were able to adapt the proofs to the order of magnitude of the involved parameters. In this way most arguments were based on no more than careful, straightforward approximations.

3. In the proof of lemma 1.5 and in part 2 of the proof of proposition 1.6 (ii) we were able to ignore initial irregularities in the behaviour of the coefficients c_n. For instance, we replaced $\sum_{n=1}^{\omega} \log c_{n+1}/c_n$ by $\sum_{n=\nu}^{\omega} \log c_{n+1}/c_n$, where ν was unlimited, but very small as regards to ω. As a consequence we had only to sum over a set of indices where the behaviour of the coefficients had stabilized, i.e. $c_{n+1}/c_n = 1+\frac{a}{n}+\frac{\emptyset}{n}$.

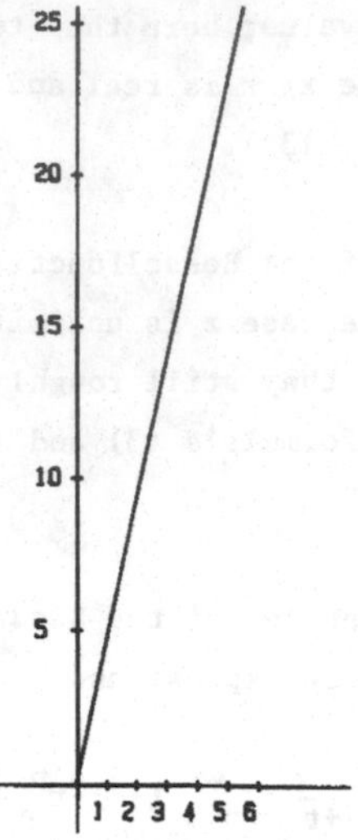

x	$\frac{\Gamma(\omega+x)}{\Gamma(\omega)}$
0.5	9.99
1	100
1.5	1003.74
2	10110
2.5	101880.07
3	1030200
3.5	10442707.26
4	106110600
4.5	1080820201.27
5	11035502400

Figure 16: If ω is unlimited, then $\Gamma(\omega+x)/\Gamma(\omega) = (1+\emptyset)\omega^x$ as long as x is limited. The picture concerns the case $\omega = 100$. On the left $\log(\Gamma(\omega+x)/\Gamma(\omega))$ is drawn. The graph is linear to the naked eye. The right half of the picture shows the numerical value of $\Gamma(\omega+x)/\Gamma(\omega)$ for x=0.5 to x=5.

Again it was convenient to have at one's disposal different orders of magnitudes within $\mathbb{R}$. Notice that an easy justification of our approach was provided by a permanence principle (Robinson's sequential lemma).

4. The Hankel function $H_p^{(1)}$ is defined by an integral transformation, applied to the function $f(t) = \frac{1}{(1+t)^{p-\frac{1}{2}}}$

$$H_p^{(1)}(z) = \sqrt{\frac{2}{\pi z}}\ \frac{e^{i\Theta}}{\Gamma(p+\frac{1}{2})}\ {}_0\!\int^{\infty} e^{-t}t^{p-\frac{1}{2}}f(\tfrac{it}{2z})\,dt$$

The terms of its expansion in $1/z$ are obtained from the same integral transformation, applied to the terms of the expansion in t of f(t), and the same is true for the remainders:

$$T_k^{(1)}(z) = \sqrt{\frac{2}{\pi z}}\ \frac{e^{i\Theta}}{\Gamma(p+\frac{1}{2})}\ {}_0\!\int^{\infty} e^{-t}t^{p-\frac{1}{2}}t_k(\tfrac{it}{2z})\,dt$$

$$R_{n-1}^{(1)}(z) = \sqrt{\frac{2}{\pi z}}\ \frac{e^{i\Theta}}{\Gamma(p+\frac{1}{2})}\ {}_0\!\int^{\infty} e^{-t}t^{p-\frac{1}{2}}r_{n-1}(\tfrac{it}{2z})\,dt$$

Part 2 of the proof of proposition 1.7(ii) shows that the integral transformation also permitted to transform an approximation of the remainder r_{n-1} into an approximation of the remainder R_{n-1}. Theorem 1.4 played a substantial role here.

So the integral transformation made it possible to transform approximations of remainders which had been obtained in a direct way in the case of convergent series (proposition 1.6) into approximations for remainders associated to divergent series.

4. The configuration of the Taylor-polynomials of the exponential function.

§1. Formulation of the problem.

The figures beneath represent the Taylor polynomials $P_n(x) = \sum_{k=0}^{n} \frac{x^k}{k!}$ of e^x, and the remainders $R_n(x) = e^x - P_n(x)$:

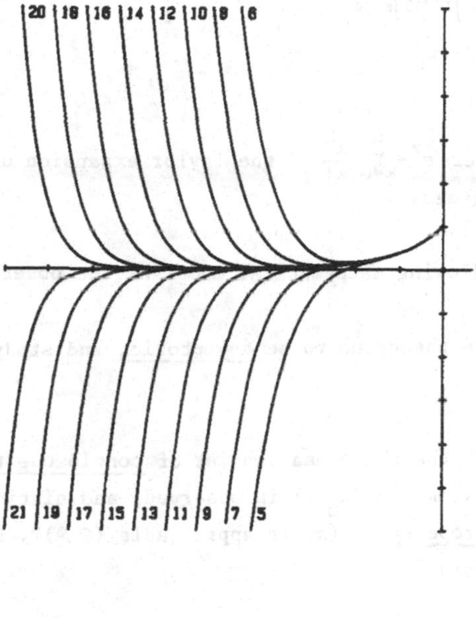

Figure 17: The exponential function e^x approached by its Taylor polynomials $\sum_{k=0}^{n} \frac{x^k}{k!}$ for $x \leqq 0$. The numbers indicate the degrees.

One observes the following particularities:

1. The graph of a Taylor polynomial closely follows the graph of the exponential function over some distance, then takes off rapidly.

2. On the domains where the graphs of the remainders are visible, and are distinct from the axis of x, they look similar, with a constant horizontal shift.

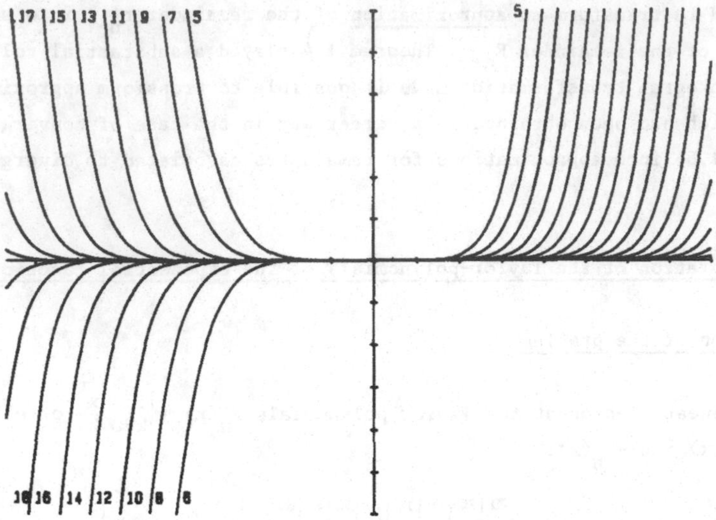

Figure 18: Remainders $e^x - \sum_{k=0}^{n} \frac{x^k}{k!}$ of the Taylor expansion of e^x. The numbers indicate the degrees.

We propose the following mathematical model of the observed phenomena:

(i) We consider the phenomena to be asymptotic, and study the behaviour of R_ω for unlimited ω

(ii) We assume that the phenomena consist of continuous transitions without sharp boundary, and formalize "$R_\omega(x)$ is observed, and distinct from the axis of x" by the external property "$R_\omega(x)$ is appreciable ($\in \mathbb{A}$)". So we try to solve the "external equation"

$$S = \{x \in \mathbb{R} \mid R_\omega(x) \in \mathbb{A}\}$$

§2. Results.

Proposition 1.9: Let ω be unlimited. Define

$$\omega' = \frac{1}{e}.\omega + \frac{1}{2e}.\log \omega$$

Then $\{x \in \mathbb{R}^+ \mid R_{\omega-1}(x) \in \mathbb{A}\} = \omega'+\mathbb{G}$. Indeed, for all limited u

$$R_{\omega-1}(\omega'+u) \simeq \frac{e}{\sqrt{2\pi}(e-1)}e^{eu}$$

Furthermore, for all limited u

$$R_{\omega}(\omega'+\frac{1}{e}+u) \simeq R_{\omega-1}(\omega'+u)$$

Figure 19: Solutions of $R_{\omega}(\omega'+d+u) = R_{\omega-1}(\omega'+u)$. Asymptotically, $d = \frac{1}{e} = .367879\ldots$

ω \ u	-1	0	1
10	0.375	0.384	.373
20	0.374	0.376	.373
30	0.372	0.374	.372
40	0.371	0.372	.371
50	0.370	0.371	.371
60	0.370	0.371	.370
70	0.369	0.370	.370
80	0.369	0.370	.370
90	0.369	0.370	.370
100	0.369	0.370	.369

Proof: Put $T_{\omega}(x) = \frac{x^{\omega}}{\omega!}$. We first solve the external equation $S = \{x \mid T_{\omega}(x) \in \mathbb{A}\}$. From Stirling's formula

$$|T_{\omega}(x)| = \frac{(1+\emptyset)}{\sqrt{2\pi\omega}}\left(\frac{xe}{\omega}\right)^{\omega}$$

By inspection we see that $T_{\omega}(\omega'+u) \simeq \frac{e^{eu}}{\sqrt{2\pi}}$ if u is limited. Hence $S = \omega'+\mathbb{G}$. We now show that $R_{\omega-1}(x)$ is appreciable iff $T_{\omega}(x)$ is appreciable. Firstly, from the expression of the remainder for Taylorexpansions in integral form we obtain for limited u, using the lemma of dominated approximation:

$$R_{\omega-1}(\omega'+u) = T_{\omega}(\omega'+u) \int_0^{\omega}(1-\frac{s}{\omega})^{\omega-1}e^{\frac{\omega'+u}{\omega}s}\,ds$$

$$\simeq \frac{e^{eu}}{\sqrt{2\pi}}\int_0^{\infty}e^{-s+s/e}\,ds \simeq \frac{e}{\sqrt{2\pi}(e-1)}e^{eu}$$

Secondly, the remainder is increasing with x. This clearly follows from the formula $R_{\omega-1}(x) = \sum_{n=\omega}^{\infty} \frac{x^n}{n!}$. Hence also $\{x \mid R_{\omega-1}(x) \in \mathbb{A}\} = \omega'+\mathbb{G}$.

Furthermore, notice that $(\omega+1)' = \frac{1}{e}(\omega+1) + \frac{1}{2e}\log(\omega+1) \simeq \frac{\omega}{e}+\frac{1}{e}+\frac{1}{2e}\log\omega = \omega'+1/e$. Hence, for limited u

$$R_{\omega}(\omega'+\frac{1}{e}+u) = R_{\omega}((\omega+1)'+u+\emptyset) \simeq \frac{e}{\sqrt{2\pi}(e-1)}e^{eu} \simeq R_{\omega-1}(\omega'+u)$$

§3. Comments on the results.

1). Figure 19 shows some numerical data concerning the horizontal shift observed on Figure 18. As we have proved above, this shift is asymptotically equal to 1/e.

2). It should be stressed that the shift-phenomenon is only local: of course two Taylor polynomials of different degree are intersecting.

3). In complex directions the shift-phenomenon is combined with a rotation phenomenon. Indeed, let Θ be some angle. One finds in the same way as above

$$R_{\omega-1}((\omega'+u)e^{i\Theta}) \simeq e^{i\Theta\omega}\frac{e}{\sqrt{2\pi}(e-e^{i\Theta})}e^{eu} \qquad \text{(u real and limited)}$$

Hence $R_{\omega}((\omega'+\frac{1}{e}+u)e^{i\Theta}) \simeq e^{i\Theta}R_{\omega-1}((\omega'+u)e^{i\Theta})$. So the approximation advances like a screwdriver: by augmenting the degree one progresses over about 1/e, and turns over an angle which nearly equals Θ. In particular, when $\Theta = \pi$, i.e. on the negative real axis, the polynomials are alternating about the function (Fig. 18, 20).

4). The observed regularity is common to many "natural" entire functions (for instance the sine and cosine function, and the Bessel functions). Indeed, it is shown in chapter II and III that if the coefficients of the Taylor series have regular asymptotic behaviour the remainders have regular asymptotic behaviour too. In particular the formula

$$R_{\omega}(x+u.\frac{x}{\omega}) = (1+\emptyset)R_{\omega}(x)e^{u} \qquad \text{(u limited)}$$

is valid for a wide range of analytic functions.

5). The problem to describe the configuration of Taylor polynomials for analytic functions at least dates back to 1936, when graphs showing the Taylor polynomials of the sine and cosine functions were published by H. Kammerer in the American Mathematical Monthly [50]. I am interested in, but do not know of other attempts to solve the problem.

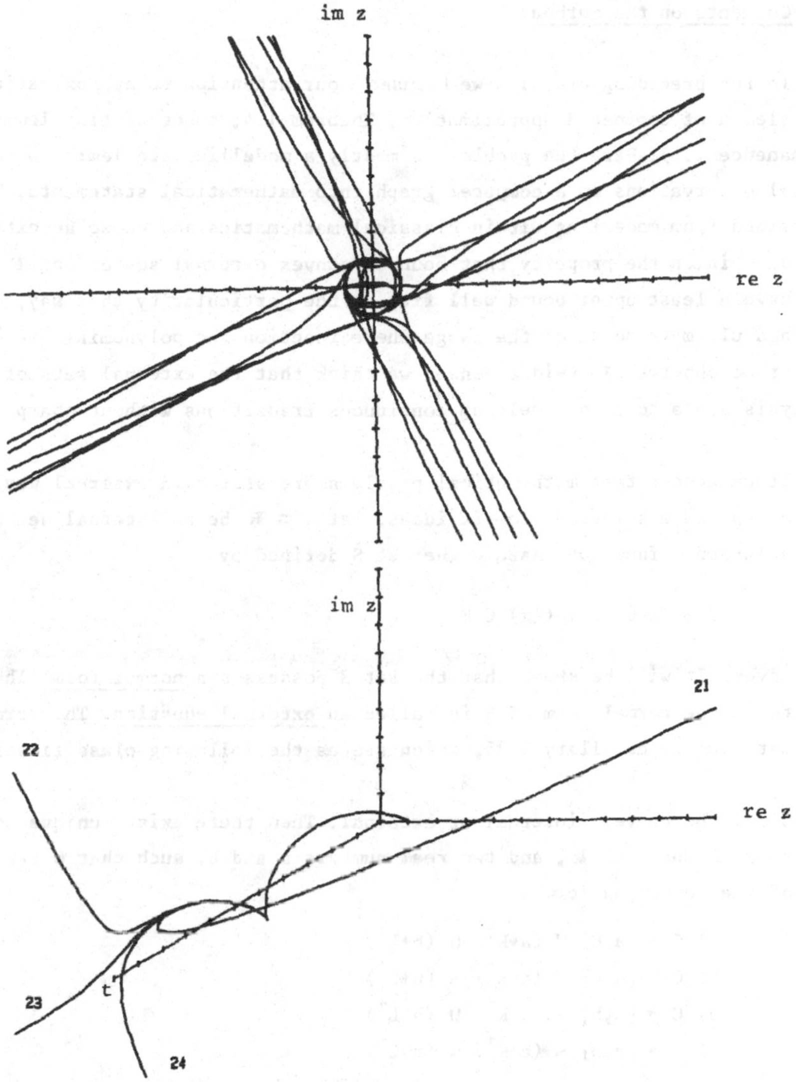

Figure 20: Taylor polynomials $\sum_{k=0}^{n} \frac{(it)^n}{n!}$ of the function $t \to e^{it}$, $t \in \mathbb{R}$. Here $\Theta = \frac{\pi}{2}$, so asymptotically the polynomials turn around the exact value with angular velocity (as a function of n) nearly equal to $\frac{\pi}{2}$. In the picture below the polynomials are drawn for n=21 to n=24 (the numbers indicate the degrees) and for t=0 upwards in the space (t, re z, im z,); in the upper picture they are drawn for n=9 to n=28, after projection on the z plane.

§4. Comments on the method.

1). In the preceding examples we focused our attention to approximation techniques (lemma of dominated approximation, Theorem 1.4, concentration lemma, rescaling, permanence ...). Here the problem is mostly a modelling problem: how to translate visual observations on a computer graph into mathematical statements. We conciously refrained from modelling within classical mathematics and chose an external model. In our opinion the property that bounded convex external subsets of $\mathbb{R}$ in general do not have a least upper bound well fits to the particularity that say, a clear, well defined ultimate point of the range where function and polynomial are confounded, cannot be observed. In a wider sense, we think that the external sets of nonstandard analysis are a tool in modelling continuous transitions without sharp boundary.

2). If we accept that mathematical problems are stated in external way, what do we accept as a solution? To fix ideas, let $E \subset \mathbb{R}$ be an external set and $f: \mathbb{R} \to \mathbb{R}$ be an internal function. Assume the set S defined by

$$S = \{x \in \mathbb{R} \mid f(x) \in E\} \tag{5}$$

is convex. It will be shown that the set S possesses a <u>normal form</u>. The problem to determine the normal form of S is called an <u>external equation</u>. The term "normal form" is justified by corollary 4.35, which states the following classification:

Let $C \subset \mathbb{R}$ be convex, internal or external. Then there exist unique convex additive subgroups K and L of $\mathbb{R}$, and two real numbers a and b, such that C takes one and only one of the following forms:

1) $C = [a,b] \cup (a+K^-) \cup (b+L^+)$
2) $C = [a,b] \cup (a+K^-) \setminus (b+L^-)$
3) $C = [a,b] \setminus (a+K^+) \cup (b+L^+)$
4) $C = [a,b] \setminus (a+K^+) \setminus (b+L^-)$

<u>Examples</u>: <u>type 1</u>: $C = \{x \in \mathbb{R} \mid R_\omega(x) \in \mathbb{A}\} = [-\omega',\omega'] \cup (-\omega'+\mathbb{G}^-) \cup (\omega'+\mathbb{G}^+)$

<u>type 2</u>: $C = \{x \in \mathbb{R}^+ \mid R_\omega(x) \simeq 0\} = [0,\omega'] \cup \{0\} \setminus \omega'+G^-$

<u>type 3</u>: $C = \{x > 0 \mid (\forall^{st} n \in \mathbb{N})(x < \varepsilon^n)$, where $\varepsilon \simeq 0\} = (\varepsilon-M)^+ \setminus \{0\}$

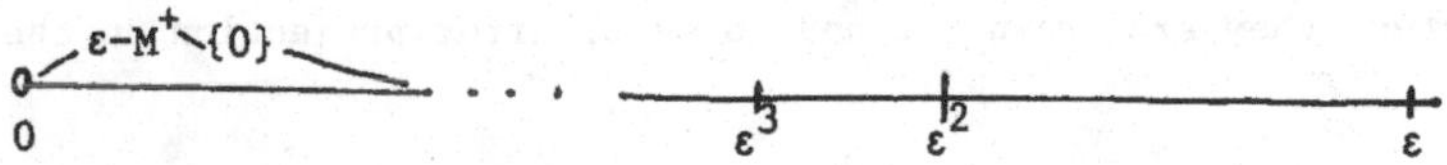

Type 4: $C \equiv \{x \in \mathbb{R}^+ \mid e^{-1/x^2} \in \varepsilon\text{-hal, where } \varepsilon \simeq 0\} = [0,(-\log \varepsilon)^{-\frac{1}{2}}] \setminus \{0\} \setminus ((-\log \varepsilon)^{-\frac{1}{2}} + ((-\log \varepsilon)^{-3/2}\text{-gal})^-)$

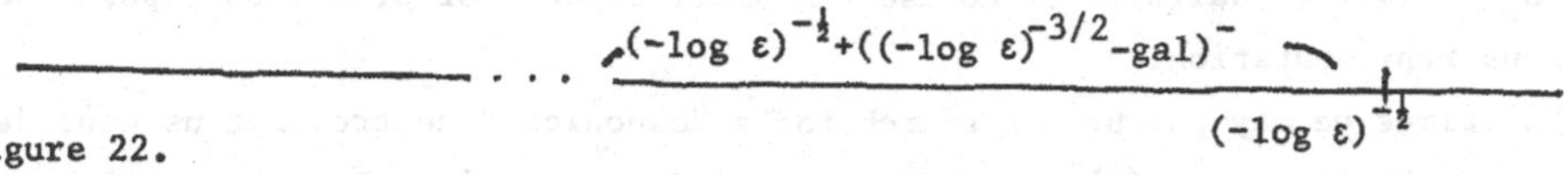

See Figure 22.

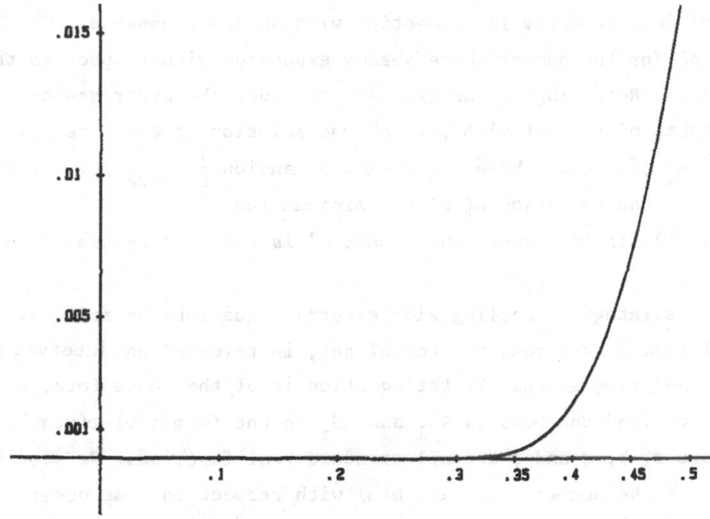

Figure 22: Graph of the function e^{-1/x^2}. If $\varepsilon \simeq 0$, a straightforward calculation shows that for $x = (-\log \varepsilon)^{-\frac{1}{2}} + u(-\log \varepsilon)^{-3/2}$ with limited u

$$e^{-1/x^2} = (1+\emptyset)\varepsilon e^{2u}$$

So e^{-1/x^2} is small with respect to ε if $x-(-\log \varepsilon)^{-\frac{1}{2}}$ is negative and not of order $(-\log \varepsilon)^{-3/2}$. If we take $\varepsilon = 0.001$, then $(-\log \varepsilon)^{-\frac{1}{2}} = 0.380....$ and $(-\log \varepsilon)^{-3/2} = 0.055...$.

As is to be seen on the above examples, the "centres" a and b of the translated subgroups are not unique.

3). In practice it may well be possible to specify further the translated subgroups and its centres. We consider the special case of a convex set C, which is solution of an external equation of the form

$$C = \{x \mid f(\omega,x) \in \mathfrak{G}\}$$

where f is a standard real function (one may replace $\mathfrak{G}$ by other parameter-free defined external sets like $\mathbb{A}$ or hal(0); the equation $C = \{x \geq 0 \mid R_\omega(x) \in \mathbb{A}\}$ belongs to this kind of equations). Let ∂S_r be the translated subgroup which represents the "right border" of C. Then on behalf of Theorem 4.40 the set ∂S_r may be written as $a(\omega)+\varphi(\omega,\mathbb{G})$ where a and φ are standard real functions, in case $\partial S_r \subset C$; in the other case, i.e. $\partial S_r \subset C^c$ then there are standard real functions b and ψ such that $\partial S_r = b(\omega)+\psi(\omega,hal(0))$. Of course the "left border" of C, say ∂S_1, possesses an analogous representation.

Sometimes we may pursue our search for a "canonical" centre. Let us consider the numbers of the form $a = a(\omega)$, where a is a standard function. It is possible that some of them have a minimum property in connection with shadow expansions. Given an order scale we may look for the number whose shadow expansion with respect to this scale has minimum length. Returning to our example, consider the order scale $(\omega,\log \omega,1)$. All elements of the set $\omega'+\mathbb{G}$, which was solution of the external equation $S = \{x \geqq 0 \mid R_{\omega-1}(x) \in \mathbb{A}\}$ have the shadow expansion $\frac{1}{e}.\omega+\frac{1}{2e}.\log \omega+c.1+\emptyset$, where c is standard. Now the expansion of ω' has minimum length $(\omega' = \frac{1}{e}.\omega+\frac{1}{2e}.\log \omega+0.1+0)$. In this sense the number ω' is the true centre of the translated group $\omega'+\mathbb{G}$.

Let us resume the strategy in dealing with external equations as follows. We first seek the normal form of the convex external set, in terms of an interval and two translated convex additive groups. If the equation is of the above form, or related forms, we try to find the borders ∂S_1 and ∂S_r in the form $a(\omega)+\varphi(\omega,\mathbb{G})$, or $b(\omega)+\psi(\omega,hal(0))$, where a, b, φ and ψ are all standard real functions. We search for shadow expansions of the numbers $a(\omega)$ and $b(\omega)$ with respect to some order scale and finally we retain the two numbers with the shortest expansion.

5. Conclusion.

The preceding sections illustrated the use of 7 important tools of nonstandard asymptotic analysis:

1) External sets and relations

2) Permanence principles

3) The classification of convex external subsets of $\mathbb{R}$

4) The lemma of dominated approximation

5) The concentration lemma

6) The Laplace method with parameters

7) Shadow expansions

In Chapters II and III we attempt to embed the examples of this chapter into a general theory of approximation. Chapters IV, V and VI are devoted to the tools.

PART II

CHAPTER II. ASYMPTOTIC EXPRESSIONS FOR THE REMAINDERS ASSOCIATED TO EXPANSIONS OF TYPE $\sum_{n=0}^{\infty} c_n \frac{z^n}{n!}$, $\sum_{n=0}^{\infty} c_n z^n$ AND $\sum_{n=0}^{\infty} c_n n! z^n$, WHERE $c_{n+p}/c_n \to c$.

The main part of this chapter concerns asymptotic expressions for the remainders of high order associated to Taylor expansions of entire and meromorph functions, and for remainders for divergent asymptotic expansions. In the latter case we restrict ourselfs to analytic functions which may be written as a Borel or Laplace transform. We suppose that the coefficients c_n have regular behaviour, i.e. $\lim_{n\to\infty} c_{n+p}/c_n = c$. At the end we present some generalizations concerning mildly irregular behaviour and other types of growth of the coefficients.

1. Definitions and notations. Some basic properties.

Definition 2.1. Suppose a number f is approximated by a partial sum $S_n = \sum_{k=0}^{n} T_k$ of terms T_k, determined by some formal expansion $(T_k)_{k\in\mathbb{N}}$. Let $R_{n-1} = f-S_{n-1}$ be the remainder after n terms. Then the ratio

$$A_n = R_{n-1}/T_n$$

will be called the approximation factor of the new term T_n with respect to the actual remainder R_{n-1}. If $f = f(x)$, we write $S_n(x)$, $T_n(x)$, $R_n(x)$ and $A_n(x)$. Occasionally, we will write $S_n^f(x)$, $T_n^f(x)$, $R_n^f(x)$ and $A_n^f(x)$.

Comments. 1. The approximation factor measures the improvement of the approximation resulting from the addition of the new term. For instance, if $A_n \simeq 1$, then the new term T_n will almost wipe out the existing remainder; if $A_n \simeq 2$ or $A_n \simeq 0$ the quality of the existing remainder remains almost unchanged. If $A_n \gtrapprox 2$ or $A_n \lessapprox 0$, the approximation surely worsens. See Fig. 1.

2. In the case of Taylor expansions there exists an explicit formula for the approximation factor. Let $f: \mathbb{R} \to \mathbb{R}$ be a C^∞ function. Then

$T_k^f(x) = \frac{f^{(k)}(0)}{k!}x^k$ is the kth term of its Taylor expansion. The formula for the remainder in integral form runs

$$R_{n-1}^f(x) = \frac{1}{(n-1)!}\int_0^x (x-t)^n f^{(n)}(t)dt$$

Hence

$$A_n^f(x) = \frac{R_{n-1}^f(x)}{T_n^f(x)} = \int_0^n (1-\frac{u}{n})^{n-1} \frac{f^{(n)}(\frac{ux}{n})}{f^{(n)}(0)}du$$

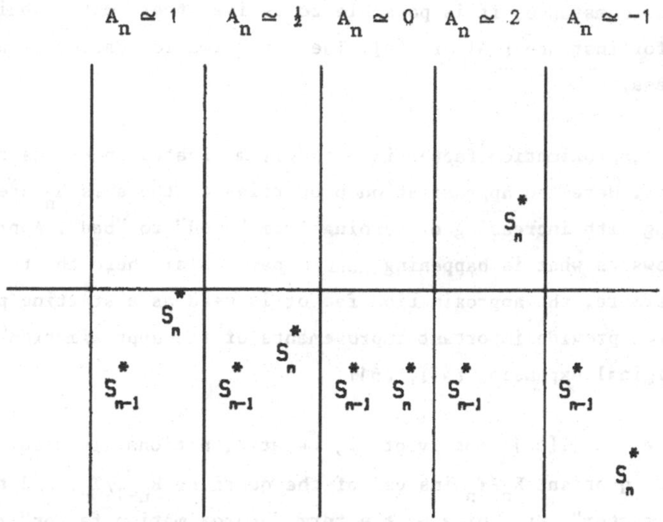

Figure 1: Geometrical interpretation of the approximation factor.

This formula can be used to obtain asymptotic expressions for the remainder of Taylor expansions. However, in this section we will use simpler formula's.

3. Again for Taylor expansions, we obtain from the formula $R_{n-1}(x) = \frac{x^n}{n!} f^{(n)}(\theta x)$, with $0 < \theta < 1$, the classical majoration of the remainder

$$(1) \qquad |R_{n-1}(x)| \leqq \left|\frac{x^n}{n!} \sup_{0 \leqq t \leqq x} f^{(n)}(t)\right| \qquad (x > 0)$$

This formula has incontestable worth as a "safety-belt" in approximations by Taylor expansions. However -and this is one of the motivations to study of the approximation factor- the majoration may be rather unrealistic. For instance, let $f(x) = e^x$. Then $T_n(x) = \frac{x^n}{n!}$ and $f^{(n)}(x) = e^x$. So (1) yields the majoration for $x \geqq 0$

$$|A_n(x)| \leqq e^x$$

Compare this with the numerical data below.

x	$A_{100}(x)$	e^x
0.1	1.00099107064	1.10517091808
1.0	1.00999901018	2.71828182846
5.0	1.05205541195	148.413159103
10.0	1.10875931174	22026.4657948
15.0	1.17407097958	3269017.37247

x	$A_{10}(x)$	e^x
1	1.099112183350	2.71828182846
2	1.217550585490	7.38905609893
3	1.360842467500	20.0855369232
4	1.538566581990	54.5981500331
5	1.755276467350	148.413159103

Figure 2: The approximation factor $A_n(x)$ associated to the Taylor expansion of e^x and its classical majoration (e^x). In the first table n=100, in the second n=10. Theoretically (see Theorem 2.14) $A_n(x) \simeq \frac{1}{1-x/n}$ if n is unlimited, and $x/n \lneqq 1$; in particular $A_n(x) \simeq 1$ if x is limited.

Of course in many circumstances it is possible to derive other, more sophisticated error bounds. See for instance [54] or [63]. The approximation factor helps us to judge their sharpness.

4. The study of the approximation factor is strongly motivated in the case of divergent expansions. Here the approximation properties of the sums S_n are necessarily changing with increasing n, turning from "good" to "bad". Approximate knowledge of A_n shows us what is happening, and in particular where the turning point is situated. Furthermore, the approximation factor is used as a starting point for new expansions, which provide important improvements of the approximation results obtained by the original expansion [43], [61].

5. In the literature ([43],[54]) mostly other, related, notions are used. Most authors consider the quotient R_n/T_n instead of the quotient R_{n-1}/T_n, and then speak of the "converging factor". My choice of the term "approximation factor" is motivated by the wish to keep a clear distinction between the effect of approximation of a number, and the ultimate convergence or divergence of the formal development inducing the approximation.

6. In section 2 and Chapter III we present asymptotic formula's for $A_n(z)$ in the case of a fairly general class of functions. It seems that results in such a general context are new. The class includes many special functions. Classically, asymptotic formula's of the approximation factors for special functions are mostly given in the form $A_{n(z)}(z)$ (one parameter case); for the two parameter case $A_n(z)$ only formal derivations are given. In this context our contribution consists of rigorous proofs for the two-parameter case. Our proofs use in essential way the possibility to relate the parameters in external way, by their order of magnitude.

<u>Definition 2.2</u>: Let f and g be two real or complex functions. We write

$$f(x) = O(g(x)) \text{ for } x \to 0$$

if there exist constants A,B > 0 such that $|f(x)| \leq A|g(x)|$ for $|x| \leq B$. We write

$$f(x) = o(g(x)) \text{ for } x \to 0$$

if $\lim_{x \to 0} f(x)/g(x) = 0$.

The symbol O(.) is known as "Landau's big O symbol" and the symbol o(.) is known as "Landau's small o symbol". In the same spirit we define:

<u>Definition 2.3</u>: Let a and b be two real or complex numbers. We write

$$a = \pounds\, b$$

if a/b is limited and

$$a = \emptyset\, b$$

if $a/b \simeq 0$.

Convention: If the symbol "$\emptyset$" is used more than once, it may designate different infinitesimals. This is in agreement with the classical use of the symbol "o(1)". Likewise, the symbol "£" may designate different limited numbers. The symbol £ was introduced by Callot, Goze and Lutz (private communication). The symbol $\emptyset$ was introduced in [13].

Definition 2.4: Let f be a real or complex function. The function f is said to possess an asymptotic expansion $\sum_{n=0} c_n x^n$ in 0 if there exists a sequence of constants $(c_n)_{n\in\mathbb{N}}$ such that for each n

$$f(x)-\sum_{k=0}^{n} c_k x^k = o(x^n)$$

for $x \to 0$. The constants are determined by induction:

$$c_n = \lim_{x\to 0} \Big(f(x)-\sum_{k=0}^{n-1} c_k x^k\Big)/x^n$$

If $\sum_{n=0} c_n x^n$ is the asymptotic expansion of f in 0, we write

$$f(x) \sim \sum_{n=0} c_n x^n$$

The definition of asymptotic expansions stems from Poincaré ([56] 1886). Here we only mention some basic properties. The asymptotic expansion of a function, if it exists, is unique, but two different functions may have the same asymptotic expansion, like e^{-1/x^2} and 0. The series $\sum_{n=0} c_n x^n$ may be convergent or divergent. Below we prove the well-known properties that the Taylor expansion is an asymptotic expansion (proposition 2.6) and that the Borel-transformation transforms asymptotic expansions into asymptotic expansions (Theorem 2.9). We start with a nonstandard characterization of the introduced notions.

Proposition 2.5 (Nonstandard characterization of O(.), o(.) and asymptotic expansions). Let f and g be standard functions

(i) $f(x) = O(g(x))$ for $x \to 0$ iff $f(\varepsilon) = £.g(\varepsilon)$ for all $\varepsilon \simeq 0$

(ii) $f(x) = o(g(x))$ for $x \to 0$ iff $f(\varepsilon) = \emptyset.g(\varepsilon)$ for all $\varepsilon \simeq 0$

(iii) $f(x) \sim \sum_{n=0} c_n x^n$ iff $c_n = {}^{o}\big(f(\varepsilon)-\sum_{k=0}^{n-1} c_k \varepsilon^k\big)/\varepsilon^n$ for all st n and $\varepsilon \simeq 0$.

Proof (i) (⇒) Let A, B > 0 be constants such that $|f(x)| \leq A|g(x)|$ for $|x| \leq B$. By the Transfer principle, they may be supposed standard. Hence $f(\varepsilon) = \pounds . g(\varepsilon)$ for all $\varepsilon \simeq 0$.

(⇐) For all $\varepsilon \simeq 0$ there is a limited number $L(\varepsilon)$ such that $|f(\varepsilon)| \leq L(\varepsilon)|g(\varepsilon)|$. Now every positive unlimited A and strictly positive infinitesimal B will do.

(ii) The assertion follows immediately from the nonstandard characterization of the limit: $\lim_{x \to 0} f(x)/g(x) = 0$ iff $f(\varepsilon)/g(\varepsilon) \simeq 0$ for all $\varepsilon \simeq 0$.

(iii)(⇒)By external induction, we see that c_n is standard if n is standard. Now for all $\varepsilon \simeq 0$ one has $f(\varepsilon)-\sum_{k=0}^{n} c_k\varepsilon^k = \emptyset.\varepsilon^n$. Hence

$$^{o}(f(\varepsilon)-\sum_{k=0}^{n-1} c_k\varepsilon^k)/\varepsilon^n) = {}^{o}(c_n+\emptyset) = c_n$$

(⇐)Let st n. We have $f(\varepsilon)-\sum_{k=0}^{n} c_k\varepsilon^k = (c_{n+1}+\emptyset)\varepsilon^{n+1} = \emptyset.\varepsilon^n$, for all $\varepsilon \simeq 0$. Hence $f(x)-\sum_{k=0}^{n} c_k x^k = o(x^n)$ for all st n, and by transfer, for all $n \in \mathbb{N}$.

In Chapter VI we use this nonstandard characterization to define an external notion "shadow expansion of numbers" which is proposed as a nonstandard alternative to the classical notion "asymptotic expansion of functions".

Proposition 2.6: Let f be a real or complex C^∞ function. Then its Taylor expansion $\sum_{n=0}^{\infty} \frac{f^{(n)}(0)}{n!}x^n$ is an asymptotic expansion.

Proof. By Transfer, we may suppose f and n standard. Then, using the fact that $f^{(n)}$ is a standard continuous function, we obtain for $\varepsilon \simeq 0$

$$f(\varepsilon)-\sum_{k=0}^{n} \frac{f^{(k)}(0)}{k!}\varepsilon^k = \frac{f^{(n+1)}(\Theta\varepsilon)}{(n+1)!}\varepsilon^{n+1} = (1+\emptyset)\frac{f^{(n+1)}(0)}{(n+1)!}\varepsilon.\varepsilon^n = \emptyset.\varepsilon^n$$

Hence $\lim_{x \to 0} (f(x)-\sum_{k=0}^{n} \frac{f^{(k)}(0)}{k!}x^k)/x^n = 0$, as required.

The addition of a new term to a Taylorpolynomial of low order effectuates a sensible improvement of the approximation result near the centre of expansion. Let us express this well-known empirical property of Taylorexpansions in the following way:

Corollary 2.7: Let f be standard C^∞, let n be limited and $\varepsilon \simeq 0$. If $T_n(\varepsilon) \neq 0$, then $A_n(\varepsilon) \simeq 1$.

n	e^x	e^{-x}	$_0\int^\infty \frac{e^{-t}}{1+xt}dt$
0	1.000500167E-003	-9.995001666E-004	-9.980059781E-004
1	5.001667083E-007	4.998333750E-007	1.994023881E-006
2	1.667083417E-010	-1.666250083E-010	-5.976119265E-009
3	4.167500139E-014	4.165833472E-014	2.388071500E-011
4	8.334722421E-018	-8.331944643E-018	-1.192850000E-013
5	1.389087326E-021	1.388690501E-021	7.149999607E-016
6	1.984375028E-025	-1.983878986E-025	-5.000039291E-018
7	2.480434331E-029	2.479883165E-029	3.996070936E-020
8	2.756007521E-033	-2.755456374E-033	-3.592906438E-022
9	2.755982464E-037	2.755481422E-037	3.589356060E-024
10	2.505419622E-041	-2.505002087E-041	-3.944383953E-026
11	2.087836301E-045	2.087515120E-045	4.728604704E-028
12	1.606019098E-049	-1.605769684E-049	-6.141129610E-030
13	1.147151036E-053	1.146998093E-053	8.589119022E-032
14	7.647641708E-058	-7.646685812E-058	-1.287100982E-033
15	4.779758494E-062	4.779196202E-062	2.057338557E-035
16	2.811613455E-066	-2.811301071E-066	-3.494043166E-037
17	1.562002907E-070	1.561838485E-070	6.283111515E-039
18	8.221048298E-075	-8.220224234E-075	-1.192621906E-040
19	4.110513362E-079	4.110121903E-079	2.382909820E-042
20	1.957383078E-083	-1.957205142E-083	-4.989218813E-044

Figure 3: Near the point of expansion, the remainders of Taylorseries decrease rapidly as a function of the index, as long as the index assumes relatively low values. This property is independent of the ultimate convergence or divergence of the series, as attested by the above table. It shows the remainders $R_n(x)$ for a convergent expansion ($e^{\pm x} = \sum_{k=0}^{\infty} \frac{(\pm x)^k}{k!}$) and a divergent expansion ($_0\int^\infty \frac{e^{-t}}{1+xt}dt \sim \sum_{k=0}(-1)^k k!.x^k$). The number x has the value 0.001.

Definition 2.8: (i) A function g is said to be of exponential order in some direction of the complex plane if there exist constants K and A such that $|g(z)| \leqq Ke^{A|z|}$ for all z in this direction. If these constants can be taken standard, then g is said to be of S-exponential order (in particular, standard functions of exponential order in standard directions are of S-exponential order in these directions).

(ii) Let g be a complex function. Then the function G(z)

$$G(z) = {_0\int^\infty} e^{-t} g(zt)dt$$

is called the (complete) Borel-transform of g. Note that if g is of exponential order then G(z) is defined for at least some z ≠ 0.

The Borel-transform is related to the Laplace-transform

$$L_g(z) \equiv {}_0\int^{\infty} e^{-zt} g(t)dt$$

by the formula

$$L_g(z) = \frac{1}{z}.G(\frac{1}{z})$$

Suppose g has an asymptotic expansion in 0, say $g(z) \sim \sum_{n=0} c_n z^n$. Under fairly general conditions its Borel-transform has also an asymptotic expansion in 0, namely $G(z) \sim \sum_{n=0} c^n n!.z^n$. This is Watson's lemma:

Theorem 2.9 (Watson's lemma): Suppose g has an asymptotic expansion $\sum_{n=0} c_n z^n$ and is of exponential order in a direction Θ of the complex plane. Then $G(z) = {}_0\int^{\infty} e^{-zt} g(t)dt$ has the asymptotic expansion $\sum_{n=0} c_n n! z^n$ in this direction.

Proof: By Transfer, we may suppose g and Θ to be standard. Put $t_n = c_n z^n$, $r_n(z) = g(z) - \sum_{k=0}^{n} t_k(z)$, $T_n(z) = c_n n!.z^n$ and $R_n(z) = G(z) - \sum_{k=0}^{n} T_k(z)$. Note that T_n is the Borel-transform of t_n and R_n is the Borel-transform of r_n. By proposition 2.5, it suffices to show that $R_{n-1}(\varepsilon)/\varepsilon^n \simeq c_n n!$, for st n and $\varepsilon \simeq 0$, knowing that $r_{n-1}(z)/z^n \simeq c_n$ for st n and $z \simeq 0$. Let us write

$$R_{n-1}(\varepsilon) = \varepsilon^n \, {}_0\int^{\infty} e^{-t} \frac{r_{n-1}(\varepsilon t)}{\varepsilon^n} dt$$

Put $I(t) = e^{-t} \frac{r_{n-1}(\varepsilon t)}{\varepsilon^n}$. We approximate ${}_0\int^{\infty} I(t)dt$ with the help of the lemma of dominated approximation 5.2. If $\varepsilon t \simeq 0$, then

$$I(t) = e^{-t}(c_n + \emptyset)t^n$$

Hence $I(t) \simeq c_n e^{-t} t^n$ for limited t and $|I(t)| \leqq e^{-t}(|c_n|+1)t^n$ for all t such that $\varepsilon t \simeq 0$. If $\varepsilon t \not\simeq 0$, then $e^{-t/2}/\varepsilon^n \simeq 0$. Because g is of S-exponential order in the direction Θ, there exist st L, A such that

$$|I(t)| \leqq \left|\frac{e^{-t/2}}{\varepsilon^n}.e^{-t/2}(f(\varepsilon t) - \sum_{k=0}^{n-1} c_k(\varepsilon t)^k)\right| \leqq e^{-t/2}.L.e^{A\varepsilon t} \leqq Le^{-t/4}$$

So $|I(t)| \leqq \max((|c_n|+1)e^{-t}t^n, Le^{-t/4})$, a standard integrable function. Consequently

$${}_0\int^{\infty} I(t)dt = {}_0\int^{\infty} c_n e^{-t} t^n dt = c_n n!$$

Hence $R_{n-1}(\varepsilon)/\varepsilon^n \simeq c_n n!$, as required.

Remark: Watson's lemma is easily generalized to integrals of type

$$G(z) = {}_0\int^{\infty} e^{-zt} t^r g(t)dt \qquad (r > -1)$$

which often occur in practice. Then the asymptotic expansion $g(x) \sim \sum_{n=0} c_n z^n$ is transformed into

$$G(z) \sim \sum_{n=0} c_n \Gamma(n+r+1) z^n$$

The following notations will be used throughout this and the next chapter.

Notation 2.10: Let $(c_n)_{n\in\mathbb{N}}$ be a sequence of real numbers. The sequence will always be supposed standard, and such that $\sum_{n=0}^{\infty} c_n z^n$ possesses a nonzero radius of convergence. Let $\varphi(z)$ be the entire function

$$\varphi(z) = \sum_{k=0}^{\infty} \frac{c_k}{k!} z^k$$

We let $\tau_n(z) = \frac{c_n}{n!} z^n$ be the general term of the Taylorseries, let $\sigma_n = \sum_{k=0}^{n} \frac{c_k}{k!} z^k$ be its partial sum and $\rho_n(z) = \varphi(z) - \sigma_n(z)$ be the remainder. Let $f(z)$ be the analytic continuation of the series $\sum_{k=0}^{\infty} c_k z^k$, such that inside the convergence circle

$$f(z) = \sum_{k=0}^{\infty} c_k z^k$$

We let $t_n(z) = c_n z^n$ be the general term of this Taylorseries, let $s_n(z) = \sum_{k=0}^{n} c_k z^k$ be its partial sum and $r_n(z) = f(z) - s_n(z)$ be the remainder. Note that one has the representation

$$f(z) = {}_0\int^{\infty} e^{-t} \varphi(zt)dt$$

I.e., if the integral is uniformly convergent on the segment joining 0 and $\bar{z}$ the function $f(\bar{z})$ and the Borel transform ${}_0\int^{\infty} e^{-t} \varphi(\bar{z}t)dt$ coincide. We again define a Borel transform. Let

$$F(z) = {}_0\int^{\infty} e^{-t} f(zt)dt$$

whenever this integral is convergent. If f is of exponential order in some direction, then $F(z)$ is defined for at least some $z \neq 0$ in this direction. In such directions the function F possesses the divergent asymptotic expansion.

$$F(z) \sim \sum_{k=0}^{\infty} c_k k! z^k$$

We let $T_k(z) = c_k k! z^k$ be the general term, let $S_n(z) = \sum_{k=0}^{n} c_k k! z^k$ be its partial and $R_n(z) = F(z) - S_n(z)$ be the remainder. Note that T_k is the Borel transform

t_k and that R_n is the Borel transform of r_n.

2. The case $c_{n+p}/c_n \to c$

We first show that the condition $\lim_{n\to\infty} c_{n+p}/c_n = c$ is rather stable, and that the coefficients of a wide class of Taylor series satisfy the condition. Under this condition asymptotic formulae for the remainders $\rho_\omega(z)$ and $r_\omega(z)$ are derived. They are valid for unlimited ω, and in nearly the whole complex plane. We then turn to the divergent expansion $\sum_{n=0} c_n n!\,\varepsilon^n$, where $\varepsilon \simeq 0$, and derive an asymptotic expression for the remainders $R_\omega(\varepsilon)$. This expression only fails for indices close to the index of the smallest term. Yet we will be able to conclude that just for those indices optimal approximation is attained.

Stability of the condition

We show the stability of the condition $c_{n+1}/c_n \to c$ under multiplication by power series.

Theorem 2.11: Let $f(z) = \sum_{n=0}^{\infty} a_n z^n$ and $g(z) = \sum_{n=0}^{\infty} b_n z^n$ be two power series such that $\lim_{n\to\infty} a_{n+1}/a_n = a \neq 0$ and g has convergence radius $R > |1/a|$. Let $\sum_{n=0}^{\infty} c_n z^n$ be the power series of f.g. Then

(i) $\lim_{n\to\infty} c_n/a_n = g(1/a)$

(ii) if $g(1/a) \neq 0$, then $\lim_{n\to\infty} c_{n+1}/c_n = a$

Proof: By Transfer, the series f and g may be supposed standard. Let $\omega \in \mathbb{N}$ be unlimited. One has the identity

$$c_\omega = a_\omega\left(b_0 + b_1\frac{a_{\omega-1}}{a_\omega} + \ldots + a_1\frac{b_{\omega-1}}{a_\omega} + a_0\frac{b_\omega}{a_\omega}\right)$$

Write $d_n = b_n \frac{a_{\omega-n}}{a_\omega}$. The following estimations hold for all unlimited ν:

$$|a_\nu|^{1/\nu} = (1+\emptyset)a \qquad |b_\nu|^{1/\nu} \leqq (1+\emptyset)/R \qquad \frac{a_\nu}{a_\omega} = \left(\frac{1+\emptyset}{a}\right)^{\omega-\nu}$$

We first show that the last terms contribute very little to the sum. Indeed, if $\omega-n$ is limited, then

$$|d_n| = \left|a_{\omega-n}\cdot\frac{a_n}{a_\omega}\cdot\frac{b_n}{a_n}\right| = \text{£}.\text{£}.\left(\frac{1+\emptyset}{aR}\right)^n \simeq 0$$

Hence $\sum_{n=\nu+1}^{\omega} d_n \simeq 0$ whenever $\omega-\nu$ is limited. By Robinson's lemma there exists $\nu \in \mathbb{N}$ such that $\omega-\nu$ is unlimited, and still $\sum_{n=\nu+1}^{\omega} d_n \simeq 0$. Secondly, we approximate $\sum_{n=0}^{\nu} d_n$. One has $d_n \simeq b_n/a^n$ for limited n and $|d_n| = \pounds(\frac{1+Ra}{2aR})^n$ for all $n \leq \nu$. So by the lemma of dominated approximation $\sum_{n=0}^{\nu} d_n \simeq \sum_{n=0}^{\infty} b_n a^n = g(1/a)$. Combining, $\sum_{n=0}^{\omega} d_n \simeq g(1/a)$.

(ii) The result follows immediately from (i).

Examples: Elementary functions like $(1+z)^r$, with $r \notin \mathbb{N}$, $\log(1+z)$ and $\Gamma(1+z)$ all have a unique singularity (in -1) on the convergence circle of their Taylor series. It is easy to verify directly that the coefficients c_n of these series have the property $\lim_{n\to\infty} c_{n+1}/c_n = -1$ (in the case of the Gamma function this is done in example 4 of chapter V). So the coefficients of the product of one of these functions with an analytic function with convergence radius greater than 1 have the same property; for instance

$$\frac{\sin z}{1+z},\ \operatorname{arctg}(\tfrac{z}{2}).\log(1+z),\ z^2e^{-z}\Gamma(1+z)\ \ldots.$$

See also Fig. 4. Notice that part (ii) of the theorem rightly excludes cases like $(1+z^2)/(1+z)$.

The above theorem is related to the classical "Theorem of Darboux" which states that the behaviour of the coefficients of a Taylor expansion about some point is roughly only influenced by the nearest singularity to the point. See for instance Henrici [47].

We now turn to the situation where the sequence of coefficients has regularily spaced zeroes, as in the case of Taylor series of functions like $\frac{1}{1+z^2}$, arcsin z or arctg z. We state a theorem which is similar to Theorem 2.11, but which is somewhat more complicated. Its proof is analogous.

Notation: Let $g(z) = \sum_{n=0}^{\infty} b_n z^n$ and $p \in \mathbb{N}$. We write for $0 \leq i \leq p-1$

$$g_i(z) = \sum_{n=0}^{\infty} b_{pn+i} z^{pn+i}$$

Theorem 2.12: Let $f(z) = \sum_{n=0}^{\infty} a_n z^n$ and $p,q \in \mathbb{N}$. Suppose there exist a sequence $(pk+q)_{k\in\mathbb{N}}$ such that $\lim_{k\to\infty} c_{p(k+1)+q}/c_{pk+q} = a$, where $a \neq 0$, and a number n_0 such that $c_n = 0$ for all indices $n \geq n_0$ not of the form pk+q. Let $g(z) = \sum_{n=0}^{\infty} b_n z^n$ be a power series with convergence radius $R > |1/a|$, and $\sum_{n=0}^{\infty} c_n z^n$ be the power series of f.g. Then for $1 \leq i \leq p-1$

(i) $\lim_{k\to\infty} c_{pk+q+i}/a_{pk+q} = g_i(1/a)$

(ii) if $g_i(1/a) \neq 0$, then $\lim_{k\to\infty} \frac{c_{p(k+1)+q+i}}{c_{pk+q+i}} = a$

<u>Figure 4</u>: If a power series $f(x) = \sum_{n=0}^{\infty} a_n x^n$, where $a_{n+1}/a_n \to a$, is multiplied by a power series $g(x) = \sum_{n=0}^{\infty} b_n x^n$ with larger convergence radius ($> |1/a|$), the asymptotic behaviour of the coefficients of the resulting series $h(x) = \sum_{n=0}^{\infty} c_n x^n$ does not change very much. Indeed, still $c_{n+1}/c_n \to a$, a consequence of the formula $c_n/a_n \to g(1/a)$ (Theorem 2.11). Here we show numerical data concerning this phenomenon.

n	c_n/a_n	c_{n+1}/c_n
0	1.00000000000	2.00000000000
1	2.00000000000	1.25000000000
2	2.50000000000	1.06666666667
3	2.66666666667	1.01562500000
4	2.70833333334	1.00307692308
5	2.71666666667	1.00051124744
6	2.71805555556	1.00007299803
7	2.71825396826	1.00000912409
8	2.71827876985	1.00000101378
9	2.71828152558	1.00000010138
10	2.71828180115	1.00000000922
11	2.71828182620	1.00000000077
12	2.71828182829	1.00000000006
13	2.71828182845	1.00000000000
14	2.71828182846	1.00000000000

$f(x) = \frac{1}{1-x} = \sum_{n=0}^{\infty} x^n$ $\qquad g(1) = e$

$g(x) = e^x$ $\qquad c_{n+1}/c_n \to 1$

$h(x) = \frac{e^x}{1-x}$

n	c_n/a_n	c_{n+1}/c_n
10	-0.96791	-0.91461
20	-0.99285	-0.95306
30	-0.99697	-0.96794
40	-0.99834	-0.97569
50	-0.99895	-0.98043
60	-0.99928	-0.98363
70	-0.99947	-0.98593
80	-0.99960	-0.98766
90	-0.99969	-0.98902
100	-0.99975	-0.99010

$f(x) = \log(1+x) = \sum_{n=1}^{\infty} \frac{(-1)^{n-1}x^n}{n}$

$g(x) = \sin \frac{\pi}{2}x$

$h(x) = \log(1+x).\sin \frac{\pi}{2}x$

$g(-1) = -1$

$c_{n+1}/c_n \to -1$

n	c_n/a_n	c_{n+1}/c_n
10	1.665	-0.852
20	1.566	-0.926
30	1.541	-0.951
40	1.530	-0.963
50	1.524	-0.970
60	1.520	-0.975
70	1.517	-0.979
80	1.515	-0.981
90	1.513	-0.983
100	1.512	-0.985

$f(x) = \sqrt{1+x} = \sum_{n=0}^{\infty} \frac{(-1)^n(-\frac{1}{2})_n}{n!} x^n$

$g(x) = \frac{1}{1+x/3}$

$h(x) = \frac{\sqrt{1+x}}{1+x/3}$

$g(-1) = \frac{3}{2}$

$c_{n+1}/c_n \to -1$

Remark: From now on we only consider the case $c_{n+p}/c_n \to -1$. Of course, this is not a proper limitation, as far as alternating series are concerned. It will often be necessary to exclude the negative real axis (which corresponds to the case $c_{n+p}/c_n = +1$).

Lemma 2.13: Let $\tau_n(z) = \frac{c_n z^n}{n!}$, $t_n(x) = c_n z^n$ and $T_n(z) = c_n n! z^n$, where $\lim_{n\to\infty} c_{n+p}/c_n = -1$. The following estimations hold for unlimited ω:

(i) $$|\tau_\omega(z)| = \left(\frac{|z|(e+\emptyset)}{\omega}\right)^\omega$$

(ii) $$|t_\omega(z)| = (|z|(1+\emptyset))^\omega$$

(iii) $$|T_\omega(z)| = e^{-\omega|z|(1-\log|\omega z|+\emptyset)/|z|}$$

Proof: The estimations follow easily from Stirling's formula and lemma 5.10.

Asymptotic formula for the remainders associated to holomorph functions.

Theorem 2.14: Let $\varphi(z) = \sum_{n=0}^{\infty} \frac{c_n z^n}{n!}$, let $\lim_{n\to\infty} \frac{c_{n+1}}{c_n} = -1$ and $\omega \in \mathbb{N}$ be unlimited. Then

(i) if $|z/\omega| \lessapprox 1$, then $\rho_{\omega-1}(z) = (1+\emptyset)\frac{\tau_\omega(z)}{1+z/\omega}$

(ii) if $|z/\omega| \gtrapprox 1$; then $\rho_{\omega-1}(z) = (1+\emptyset)\frac{\tau_\omega(z)}{1+z/\omega} + \varphi(z)$

Proof: (i) Let $|z/\omega| \lessapprox 1$. By the convergence of the series

$$\rho_{\omega-1}(z) = \sum_{n=\omega}^{\infty} \frac{c_n z^n}{n!} = \tau_\omega(z) \sum_{k=0}^{\infty} \frac{c_{\omega+k}}{c_\omega} \frac{z^k}{(\omega+1)\ldots(\omega+k)}$$

Put $u_k = \frac{c_{\omega+k}}{c_\omega} \frac{z^k}{(\omega+1)\ldots(\omega+k)}$. Then $u_k \simeq (-z/\omega)^k$ for all st k, and $|u_k| \leq \left(\frac{{}^o(|z|/\omega)+1}{2}\right)^k$ for all $k \in \mathbb{N}$. So by the lemma of dominated approximation $\sum_{k=0}^{\infty} u_k \simeq \sum_{k=0}^{\infty} (-z/\omega)^k = \frac{1}{1+z/\omega}$. Hence $\rho_{\omega-1}(z) = (1+\emptyset)\frac{\tau_\omega(z)}{1+z/\omega}$.

(ii) Let $|z/\omega| \gtrapprox 1$. One has

$$\rho_{\omega-1}(z) = \varphi(z) - \sum_{n=0}^{\omega-1} \frac{c_n z^n}{n!} = \varphi(z) - \tau_{\omega-1}(z) \sum_{k=0}^{\omega-1} \frac{c_{\omega-k-1}}{c_{\omega-1}} \cdot \frac{(\omega-1)\ldots(\omega-k)}{z^k}$$

Let $v_k = \frac{c_{\omega-k-1}}{c_{\omega-1}} \cdot \frac{(\omega-1)\ldots(\omega-k)}{z^k}$. Then $v_k \simeq (-\omega/z)^k$ for all standard k. In order to be able to apply the lemma of dominated approximation, we must eliminate the possible irregular behaviour of the v_k for st(ω-k). Now if st(ω-k), then using lemma 5.10

$$|v_{\omega-k}| \leq |c_{\omega-k-1}|e^{\emptyset.\omega}|\omega/z|^{\omega-k} \simeq 0$$

Hence $\sum_{k=\omega-m}^{\omega-1} v_k \simeq 0$ for st m. So by Robinson's lemma there exists unlimited ν such that $\sum_{k=\omega-\nu}^{\omega-1} v_k \simeq 0$. Now for $k \leq \omega-\nu$ one has $|v_k| \leq (1+\emptyset)^k(\omega/|z|)^k \leq (\frac{^o(\omega/|z|)+1}{2})^k$.
Hence, by the lemma of dominated approximation

$$\sum_{k=0}^{\omega-\nu-1} v_k \simeq \sum_{k=0}^{\infty} (\omega/z)^k = \frac{z}{\omega}\cdot\frac{1}{1+z/\omega}$$

Combining, one sees that $\sum_{k=0}^{\omega-1} v_k \simeq \frac{z}{\omega}\cdot\frac{1}{1+z/\omega}$. Hence, using the approximation

$$\frac{\tau_{\omega-1}(z)}{\tau_\omega} = -(1+\emptyset)\omega/z,$$

$$\rho_{\omega-1}(z) = \varphi(z)-\tau_{\omega-1}(z)\cdot\frac{z}{\omega}\,\frac{1+\emptyset}{1+z/\omega} = (1+\emptyset)\frac{\tau_\omega(z)}{1+z/\omega}+\varphi(z)$$

<u>Comment</u>: The appearance of the term $\varphi(z)$ in the expression of $\rho_{\omega-1}(z)$ for $|z/\omega| \gtrsim 1$ is explained by comparing orders of magnitude. Indeed, by lemma 2.13 one has the estimation

$$|\tau_\omega(z)| = (\frac{|z|(e+\emptyset)}{\omega})^\omega$$

Furthermore for real negative and unlimited z the following estimation holds

$$|\varphi(z)| = (e+\emptyset)^{|z|}$$

One could prove this as an exercise (see the exercise on p. 85 for a hint). So for real negative z such that $|z/\omega| \gtrsim 1$ the value of $\varphi(z)$ is very much higher than the value of $\left|\frac{\tau_\omega(z)}{1+z/\omega}\right|$. Of course, in other directions it may be otherwise.

The estimation of $|\tau_\omega(z)|$ shows that the approximation of $\varphi(z)$ by $\sigma_\omega(z)$ is very good for $|z/\omega| \lesssim 1/e$ and very bad for $|z/\omega| \gtrsim 1/e$.

<u>Asymptotic formula for the remainders associated to meromorph functions.</u>

<u>Theorem 2.15</u>: Let $f(z) = \sum_{n=0}^{\infty} c_n z^n$, let $\lim_{n\to\infty} \frac{c_{n+1}}{c_n} = -1$, and $\omega \in \mathbb{N}$ be unlimited. Then

(i) if $|z| \lesssim 1$, then $r_{\omega-1}(z) = (1+\emptyset)\frac{t_\omega(z)}{1+z}$

(ii) if $|z| \gtrsim 1$, then $r_{\omega-1}(z) = (1+\emptyset)\frac{t_\omega(z)}{1+z} + f(z)$.

<u>Proof</u>: (i) Let $|z| \lesssim 1$. By the convergence of the series

$$r_{\omega-1}(z) = \sum_{n=\omega}^{\infty} c_n z^n = t_\omega(z)\sum_{k=0}^{\infty} \frac{c_{\omega+k}}{c_\omega} z^k$$

In the same way as in the proof of 2.14(i) one finds the estimation $\sum_{k=0}^{\infty} \frac{c_{\omega+k}}{c_\omega} z^k \simeq \frac{1}{1+z}$. Hence $r_{\omega-1}(z) = (1+\emptyset)\frac{t_\omega(z)}{1+z}$.

(ii) Let $|z| \gtrsim 1$. One has

$$r_{\omega-1}(z) = f(z) - \sum_{n=0}^{\omega-1} c_n z^n = f(z) - t_{\omega-1}(z)\sum_{k=0}^{\omega-1} \frac{c_{\omega-k-1}}{c_{\omega-1}}(1/z)^k.$$

In the same way as in the proof of 2.14(ii) one finds the estimation $\sum_{k=0}^{\omega-1} \frac{c_{\omega-k-1}}{c_{\omega-1}}(1/z)^k \simeq \frac{z}{1+z}$. Hence, using the approximation $t_\omega(z)/t_{\omega-1}(z) = -(1+\emptyset)z$,

$$r_{\omega-1}(z) = f(z) - (1+\emptyset)t_{\omega-1}(z).\frac{z}{1+z} = (1+\emptyset)\frac{t_\omega(z)}{1+z} + f(z)$$

<u>Comments</u>: 1) For limited $|z| \gtrsim 1$, but outside the halo of any singularity of f, one may as well neglect $f(z)$ with respect to $t_\omega(z)$: the first has limited, and the second has unlimited value. So for these z the formula $r_{\omega-1}(z) = (1+\emptyset)\frac{t_\omega(z)}{1+z}$ of 2.15(i) still holds.
2) The formula $r_{\omega-1}(z) = (1+\emptyset)\frac{t_\omega(z)}{1+z}$ reflects the "almost geometrical" character of the series $\sum_{n=0}^{\infty} c_n z^n$. Indeed, in the case of the geometrical series $\sum_{n=0}^{\infty} (-1)^n z^n$ we have the exact formula $r_{n-1}(z) = \frac{t_n(z)}{1+z}$, which is valid for all n and z $(z \neq -1)$.
The asymptotic expression for $r_{\omega-1}(z)$ may very well be invalid near the convergence circle. It is well-known that the Taylor series can have very complicated behaviour in this region. See for instance [42].

<u>Asymptotic formula's for the remainders associated to divergent asymptotic expansions of complete or incomplete Borel-transforms.</u>

The two preceding theorems concerned remainders $\rho_{\omega-1}(z)$ and $r_{\omega-1}(z)$ of convergent expansions. There the unlimited index ω was fixed, and the argument z was the variable. In the next theorem the remainder $R_{n-1}(\varepsilon)$ of a divergent expansion will be investigated. Now the argument ε will be fixed, and the index n will be the variable.

<u>Theorem 2.16</u>: Let $f(z) = \sum_{n=0}^{\infty} c_n z^n$, where $\lim_{n\to\infty} c_{n+1}/c_n = -1$, and let $\varepsilon \in \mathbb{C}$ be infinitesimal, with $|\arg \varepsilon| \lesssim \pi$. Suppose f can be continuated analytically and is of S-exponential order in the direction of arg ε. Let $F(\varepsilon) = \int_0^\infty e^{-t} f(\varepsilon t)dt$. Then

(i) if $|n\varepsilon| \not\simeq 1$, then $R_{n-1}(\varepsilon) = (1+\emptyset)\frac{T_n(\varepsilon)}{1+n\varepsilon}$

(ii) if $|n\varepsilon| \simeq 1$, then $|R_{n-1}(\varepsilon)| = e^{-(1+\emptyset)/|\varepsilon|}$

<u>Proof</u>: In estimating $R_{n-1}(\varepsilon)$ we distinguish three orders of magnitude of the index: <u>A</u> st n, <u>B</u> n unlimited, $n\varepsilon$ limited, <u>C</u> $n\varepsilon$ unlimited.

<u>A st n</u>. From Watson's lemma

$$R_{n-1}(\varepsilon) = (1+\emptyset)T_n(\varepsilon)$$

Because $n\varepsilon \simeq 0$ we might as well write $R_{n-1}(\varepsilon) = (1+\emptyset)\frac{T_n(\varepsilon)}{1+n\varepsilon}$

<u>B, n unlimited, $n\varepsilon$ limited</u>. We write $n = \omega$. By the Fehrele principle there exists $\alpha \simeq 0$ such that the formula's of Theorem 2.15 are valid outside some annulus of the form $1-\alpha \leq |z| \leq 1+\alpha$, where $\alpha \simeq 0$. Now using Theorem 1.4 and Proposition 5.10

$$\begin{aligned}
R_{\omega-1}(\varepsilon) &= {}_0\int^{\infty} e^{-t} r_{\omega-1}(\varepsilon t)dt \\
&= {}_0\int^{\infty} e^{-t}\frac{t_\omega(\varepsilon t)}{1+\varepsilon t}(1+\beta(t))dt + \int_{(1-\alpha)/|\varepsilon|}^{\infty} e^{-t} f(\varepsilon t)dt \qquad (\beta(t) \simeq 0) \\
&- \sum_{k=0}^{\omega-1} \int_{(1-\alpha)/|\varepsilon|}^{(1+\alpha)/|\varepsilon|} e^{-t} t_k(\varepsilon t)dt - \int_{(1-\alpha)/|\varepsilon|}^{(1+\alpha)/|\varepsilon|} e^{-t}\frac{t_\omega(\varepsilon t)}{1+\varepsilon t}dt \\
&= \varepsilon^\omega c_\omega \, {}_0\int^{\infty} e^{-t}\frac{t^\omega}{1+\varepsilon t}(1+\beta(t))dt + \Theta_1 e^{-(1+\emptyset)/|\varepsilon|} \qquad (|\Theta_1| = 1) \\
&- \sum_{k=0}^{\omega-1} c_k\varepsilon^k \int_{(1-\alpha)/|\varepsilon|}^{(1+\alpha)/|\varepsilon|} e^{-t} t^k dt - c_\omega\varepsilon^\omega \int_{(1-\alpha)/|\varepsilon|}^{(1+\alpha)/|\varepsilon|} e^{-t}\frac{t^\omega}{1+\varepsilon t}dt \\
&= (1+\emptyset)\frac{\varepsilon^\omega c_\omega \omega!}{1+\varepsilon\omega} + \Theta_1 e^{-(1+\emptyset)/|\varepsilon|} + \Theta_2 e^{-(1+\emptyset)/|\varepsilon|} + \Theta_3 e^{-(1+\emptyset)/|\varepsilon|} \qquad (|\Theta_2|, |\Theta_3| = 1)
\end{aligned}$$

The estimations of the last three terms result from easy majorations. So we have $R_{\omega-1}(\varepsilon) = (1+\emptyset)\frac{T_\omega(\varepsilon)}{1+\varepsilon\omega} + \Theta e^{-(1+\emptyset)/|\varepsilon|}$, with $|\Theta| = 1$. Now by Lemma 2.13(iii)

$$|T_\omega(\varepsilon)| = e^{-|\omega\varepsilon|(\log|\omega\varepsilon|-1+\emptyset)/|\varepsilon|}$$

This estimation implies that $R_{\omega-1}(\varepsilon) = (1+\emptyset)\frac{T_\omega(\varepsilon)}{1+\varepsilon\omega}$ for $|\omega\varepsilon| \not\simeq 1$, and $|R_{\omega-1}(\varepsilon)| = e^{-(1+\emptyset)/|\varepsilon|}$ for $|\omega\varepsilon| \simeq 1$. Hence part (ii) is proved.

<u>C $n\varepsilon$ unlimited</u>: We write again $n = \omega$. The proof is a copy of part (ii) of the proof of proposition 1.1. Define $\nu = [3/|\varepsilon|)$. Then, using <u>B</u> and lemma 5.12:

$$\begin{aligned}
R_{\omega-1}(\varepsilon) &= R_{\nu-1}(\varepsilon) - \sum_{k=\nu}^{\omega-1} T_k(\varepsilon) = (1+\emptyset)\frac{T_\nu(\varepsilon)}{4} - T_{\omega-1}(\varepsilon)\sum_{m=1}^{\omega-\nu}\frac{T_{\omega-m}(\varepsilon)}{T_{\omega-1}(\varepsilon)} \\
&= (1+\emptyset)\frac{T_\nu(\varepsilon)}{4} - (1+\emptyset)T_{\omega-1}(\varepsilon) = -(1+\emptyset)T_{\omega-1}(\varepsilon) = (1+\emptyset)\frac{T_\omega(\varepsilon)}{\omega\varepsilon} = (1+\emptyset)\frac{T_\omega(\varepsilon)}{1+\omega\varepsilon}
\end{aligned}$$

This completes the proof of part (i).

Corollary 2.17: Both the indices of the smallest term and the smallest remainder of the divergent expansion $F(\varepsilon) \sim \sum_{n=0} c_n n! \varepsilon^n$ are of the form $(1+\emptyset)/|\varepsilon|$.

Comments: 1) By Transfer, if f is of exponential order in some standard direction, then it is of S-exponential order in this direction. There exists a direction where the property fails, namely the negative real axis.

2) We use the result to give a qualitative description of the approximation of $F(\varepsilon)$ by the sums $S_n(\varepsilon)$, like we did in the first section of chapter I. For $|n\varepsilon| \not\simeq 1$

$$\frac{R_n(\varepsilon)}{R_{n-1}(\varepsilon)} = (1+\emptyset)\frac{T_{n+1}(\varepsilon)}{T_n(\varepsilon)} = (1+\emptyset)\frac{c_{n+1}}{c_n}(n+1)\varepsilon$$

The formula implies that for st n the relative error diminishes with unlimited speed. For unlimited $n = \omega$ the formula simplifies to

$$\frac{R_\omega(\varepsilon)}{R_{\omega-1}(\varepsilon)} = -(1+\emptyset)\omega\varepsilon$$

This shows that the remainders decrease as long as $|\omega\varepsilon| \lesssim 1$, but the relative amelioration decreases almost linearly. For $|\omega\varepsilon| \simeq 1$ we know that the approximation is the best possible and is of order $e^{-(1+\emptyset)/|\varepsilon|}$ -very small indeed- but otherwise we loose sight on what is happening. For $|\omega\varepsilon| \gtrsim 1$ the remainders grow again; the relative deterioration of the approximation increases almost linearly. On the whole, this approximation process -first approaching to the exact value, then retreating from it- is fairly regular, except maybe for the indices where approximation is best, i.e. indices close to that of the smallest term.

It is often observed in practice (already by Poincaré in his Mécanique Céleste [55]) that best approximation is obtained by summing to the index of the smallest term. The above result verifies this property somewhat coarsely for complete Borel transforms. Below we show that the property is not always satisfied in the case of incomplete Borel transforms.

Definition 2.18: (i) Let $f(z) = \sum_{n=0}^{\infty} c_n z^n$. Let $a > 0$. Suppose f can be continuated analytically along the line $[0, ae^{i \operatorname{Arg} z}]$. Then the incomplete Borel transform $F_a(z)$ of f is defined by

$$F_a(z) = \int_0^{a/|z|} e^{-t} f(zt)\,dt$$

Furthermore we write

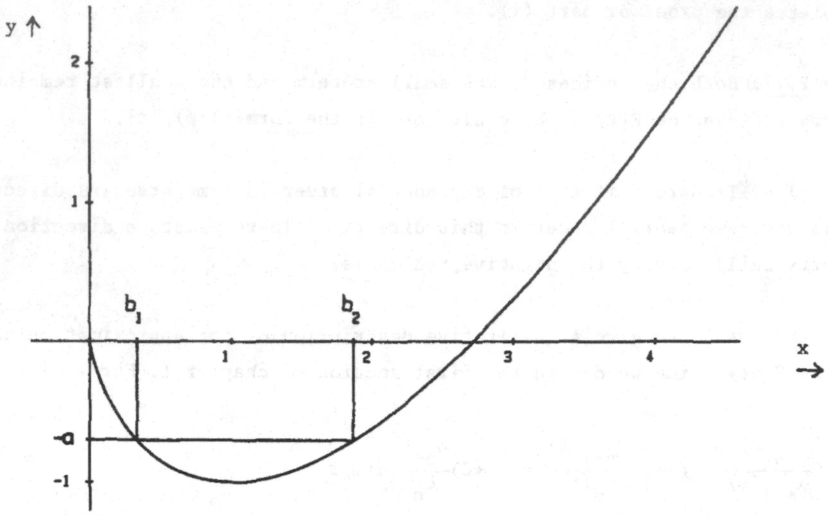

<u>Figure 5</u>: Graph of the function $x \to -x(1-\log x)$ and solutions of the equation $-x(1-\log x) = -a$, with $a = 0.7$.

$$R_n^{\,a}(z) = F_a(z) - \sum_{k=0}^{n} c_k k! z^k$$

(ii) For $0 \leqq a < 1$ the equation $-u(1-\log u) = -a$ has two solutions. The lesser one will be written b_1 and the greater one b_2. Note that if $0 \lnapprox a \lnapprox 1$, then $0 \lnapprox b_1 \lnapprox a \lnapprox 1 \lnapprox b_2 \lnapprox e$. For $a \gtrapprox 0$ the functions $F(z)$ and $F_a(z)$ have the same asymptotic expansion: $\sum_{n=0} c_n n! z^n$. This will be a consequence of the theorems below, which imply that for $\varepsilon \simeq 0$, the difference $F(\varepsilon) - F_a(\varepsilon)$ is smaller than all standard powers of ε. The first theorem shows that if a lies near or outside the convergence circle, the approximation of the numbers $F(\varepsilon)$ and $F_a(\varepsilon)$ by the partial sums of the series $\sum_{n=0} c_n n! z^n$ is very much the same. As shown by the second theorem, this is no longer true for a well inside the convergence circle.

<u>Theorem 2.19</u>: Let $f(z) = \sum_{n=0}^{\infty} c_n z^n$ where $\lim_{n \to \infty} c_{n+1}/c_n = -1$. Let $\varepsilon \in \mathbb{C}$ be infinitesimal, with $|\arg \varepsilon| \lnapprox \pi$. Suppose $a \gtrapprox 1$, and f can be continuated analytically and is of S-exponential order in the direction of $\arg \varepsilon$. Let $F_a(z)$ and $R_n^{\,a}(z)$ be as above. Then

(i) if $|n\varepsilon| \not\simeq 1$, then $R_{n-1}^{\,a}(\varepsilon) = (1+\emptyset)\dfrac{T_n(\varepsilon)}{1+n\varepsilon}$

(ii) if $|n\varepsilon| \simeq 1$, then $|R_{n-1}^{\,a}(\varepsilon)| = e^{-(1+\emptyset)/|\varepsilon|}$

Proof: We have the identity

$$R_{n-1}^{a}(\varepsilon) = R_{n-1}(\varepsilon) - \int_{a/|\varepsilon|}^{\infty} e^{-t} f(\varepsilon t)dt$$

We have the following estimates. Firstly $|R_{n-1}(\varepsilon)| = e^{-|n\varepsilon|(1-\log|n\varepsilon|+\emptyset)/|\varepsilon|}$ by Theorem 2.16 and Lemma 2.13(iii), secondly $|\int_{a/|\varepsilon|}^{\infty} e^{-t} f(\varepsilon t)| = e^{-a(1+\emptyset)/|\varepsilon|}$. These estimations and the fact that $a \gtrapprox 1$ imply that $R_{n-1}^{a}(\varepsilon) = (1+\emptyset)R_{n-1}(\varepsilon)(=(1+\emptyset)\frac{T_n(\varepsilon)}{1+n\varepsilon})$ for $|n\varepsilon| \not\simeq 1$ and $|R_{n-1}^{a}(\varepsilon)| = e^{-(1+\emptyset)/|\varepsilon|}$ for $|n\varepsilon| \simeq 1$.

Next theorem concerns the case $0 \lnapprox a \lnapprox 1$. The above proof can be adapted for this case, but then we would make an artificial assumption: we stay entirely in the convergence disk of $f(z) = \sum_{n=0}^{\infty} c_n z^n$, but ask that f can be continuated everywhere and is of S-exponential order (in fact this assumption was already artificial in Theorem 2.19; it is only needed on the linesegment$[0,ae^{i \operatorname{Arg} z}]$). Below we present a direct proof which is straightforward, but includes many estimations.

Theorem 2.20: Let $f(z) = \sum_{n=0}^{\infty} c_n z^n$ with $\lim_{n\to\infty} c_{n+1}/c_n = -1$. Let $0 \lnapprox a \lnapprox 1$, and $\varepsilon \in \mathbb{C}$, $\varepsilon \simeq 0$. Let $F_a(z)$, $R_n^{a}(z)$, b_1 and b_2 be as above. Then

(i) if $|n\varepsilon| \lnapprox b_1$ or $|n\varepsilon| \gnapprox b_2$ then $R_{n-1}^{a}(\varepsilon) = (1+\emptyset)\frac{T_n(\varepsilon)}{1+n\varepsilon}$

(ii) if $|n\varepsilon| \simeq b_1$ or $|n\varepsilon| \simeq b_2$, then $|R_{n-1}^{a}(\varepsilon)| = e^{-(a+\emptyset)/|\varepsilon|}$

(iii) if $b_1 \lnapprox |n\varepsilon| \lnapprox b_2$, then $R_{n-1}(\varepsilon) = e^{-a/|\varepsilon|}(f(ae^{i \arg \varepsilon})+\emptyset)$

Proof: Let $a \lnapprox c \lnapprox 1$. With the help of the lemma of dominated approximation one proves

$$F_a(\varepsilon)-F_c(\varepsilon) = \int_{a/|\varepsilon|}^{c/|\varepsilon|} e^{-t} f(\varepsilon t)dt = e^{-a/|\varepsilon|}(f(ae^{i \arg \varepsilon})+\emptyset)$$

Now after some algebra one gets

$$R_{n-1}^{a}(\varepsilon) = \int_0^{c/|\varepsilon|} e^{-t} r_{n-1}(\varepsilon t)dt - \sum_{k=0}^{n-1} c_k \varepsilon^k \int_{c/|\varepsilon|}^{\infty} e^{-t} t^k dt + F_a(\varepsilon) - F_c(\varepsilon)$$

Put $U_1 = \int_0^{c/|\varepsilon|} e^{-t} r_{n-1}(\varepsilon t)dt$, $U_2 = -\sum_{k=0}^{n-1} c_k \varepsilon^k \int_{c/|\varepsilon|}^{\infty} e^{-t} t^k dt$ and $U_3 = F_a(\varepsilon)-F_c(\varepsilon)$.

By Theorem 2.15 we have $U_1 = c_n \varepsilon^n \int_0^{c/|\varepsilon|} e^{-t} \frac{t^n}{1+n\varepsilon}(1+\emptyset)dt$. Below we estimate U_1 and U_2 as a function of n. We distinct several cases, taking into account the distributional character of $e^{-t}t^n$ (see Fig. 9 of Chapter I). Its peak lies at t=n and has the approximate value $(n/e)^{n(1+\emptyset)}$, if n is unlimited; generally speaking, if the peak is inside, or very close to, the interval of integration, then as a consequence of the concentration lemma (5.6) the integral has the same order of magnitude as the peak, and may again be written $(n/e)^{n(1+\emptyset)}$. If the peak is well outside, the integral has the order of magnitude of the value of the integrand at the border.

With this remark, Theorem 1.4, the lemma of dominated approximation and the estimations $U_3 = e^{-a/|\varepsilon|}f(ae^{i\arg\varepsilon}+\emptyset)$ and $|T_n(\varepsilon)|e^{-|n\varepsilon|(1-\log|n\varepsilon|+\emptyset)/|\varepsilon|}$ the justification of the following estimations is routine.

	U_1	U_2	$R_{n-1}(\varepsilon)=U_1+U_2+U_3$
n limited	$(1+\emptyset)T_n(\varepsilon)$	$\pounds e^{-c/\lvert\varepsilon\rvert}$	$(1+\emptyset)\frac{T_n(\varepsilon)}{1+n\varepsilon}$
$\lvert n\varepsilon\rvert \underset{\not\simeq}{<} b_1$ n unlimited	$(1+\emptyset)\frac{T_n(\varepsilon)}{1+n\varepsilon}$	$e^{-c/\lvert\varepsilon\rvert}(f(ce^{i\arg\varepsilon})+\emptyset)$	$(1+\emptyset)\frac{T_n(\varepsilon)}{1+n\varepsilon}$
$\lvert n\varepsilon\rvert \simeq b_1$	$(1+\emptyset)\frac{T_n(\varepsilon)}{1+n\varepsilon}=\pounds e^{-(a+\emptyset)/\lvert\varepsilon\rvert}$	$e^{-c/\lvert\varepsilon\rvert}(f(ce^{i\arg\varepsilon})+\emptyset)$	$\pounds e^{-(a+\emptyset)/\lvert\varepsilon\rvert}$
$b_1 \underset{\not\simeq}{<} \lvert n\varepsilon\rvert \underset{\not\simeq}{<} c$	$(1+\emptyset)\frac{T_n(\varepsilon)}{1+n\varepsilon}=\pounds e^{-d/\lvert\varepsilon\rvert}$ $a \underset{\not\simeq}{<} d \underset{\not\simeq}{<} 1$	$e^{-c/\lvert\varepsilon\rvert}(f(ce^{i\arg\varepsilon})+\emptyset)$	$e^{-a/\lvert\varepsilon\rvert}(f(ae^{i\arg\varepsilon})+\emptyset)$
$c \underset{\simeq}{<} \lvert n\varepsilon\rvert \underset{\not\simeq}{<} b_2$	$\pounds e^{-r/\lvert\varepsilon\rvert}$ $a \underset{\not\simeq}{<} r$	$\pounds e^{-s/\lvert\varepsilon\rvert}$ $a \underset{\not\simeq}{<} s \underset{\not\simeq}{<} 1$	$e^{-a/\lvert\varepsilon\rvert}(f(ae^{i\arg\varepsilon})+\emptyset)$
$\lvert n\varepsilon\rvert \simeq b_2$	$\emptyset . T_{n-1}(\varepsilon)$	$(1+\emptyset)\frac{T_n(\varepsilon)}{1+n\varepsilon}=\pounds e^{-(a+\emptyset)/\lvert\varepsilon\rvert}$	$\pounds e^{-(a+\emptyset)/\lvert\varepsilon\rvert}$
$\lvert n\varepsilon\rvert \underset{\not\simeq}{>} b_2$	$\emptyset . T_{n-1}(\varepsilon)$	$(1+\emptyset)\frac{T_n(\varepsilon)}{1+n\varepsilon}$	$(1+\emptyset)\frac{T_n(\varepsilon)}{1+n\varepsilon}$

The estimations of the last column prove the theorem.

As a corollary, we show that for a given formal series $\sum_{k=0} c_k k! z^k$ there exists an analytic function such that the property of "summation to the smallest term" is satisfied in the following sense:

<u>Corollary 2.21</u>: Let $\sum_{k=0} c_k k! z^k$ be a standard formal series such that $\lim_{k\to\infty} c_{k+1}/c_k = -1$. Put $T_k(z) = c_k k! z^k$. There exists an analytic function F_α such that for all infinitesimal $\varepsilon \in \mathbb{C}$ the remainders $R_{n-1}{}^{\alpha}(\varepsilon) = F_\alpha(\varepsilon) - \sum_{k=0}^{n-1} T_k(\varepsilon)$ satisfy the asymptotic expressions

$$R_{n-1}{}^{\alpha}(\varepsilon) = (1+\emptyset)\frac{T_n(\varepsilon)}{1+n\varepsilon} \qquad |n\varepsilon| \not\simeq 1$$

$$|R_{n-1}{}^{\alpha}(\varepsilon)| = e^{-(1+\emptyset)/|\varepsilon|} \qquad |n\varepsilon| \simeq 1$$

<u>Proof</u>: By the Fehrele principle there exists $\alpha \simeq 1$ such that Theorem 2.20 holds for $F_\alpha(z) = \int_0^{\alpha/|z|} e^{-t} f(zt)\,dt$, where $f(z) = \sum_{k=0}^{\infty} c_k z^k$. Then $b_1 \simeq b_2 \simeq 1$ and the conclusion follows from 2.20(i), (ii) and (iii).

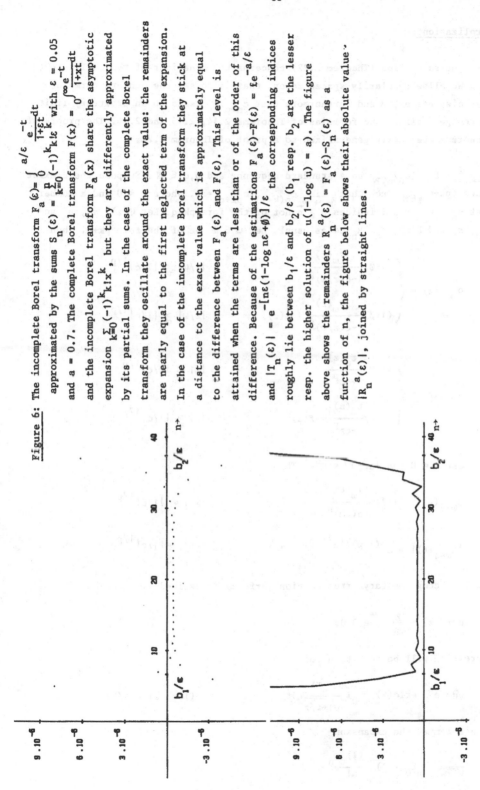

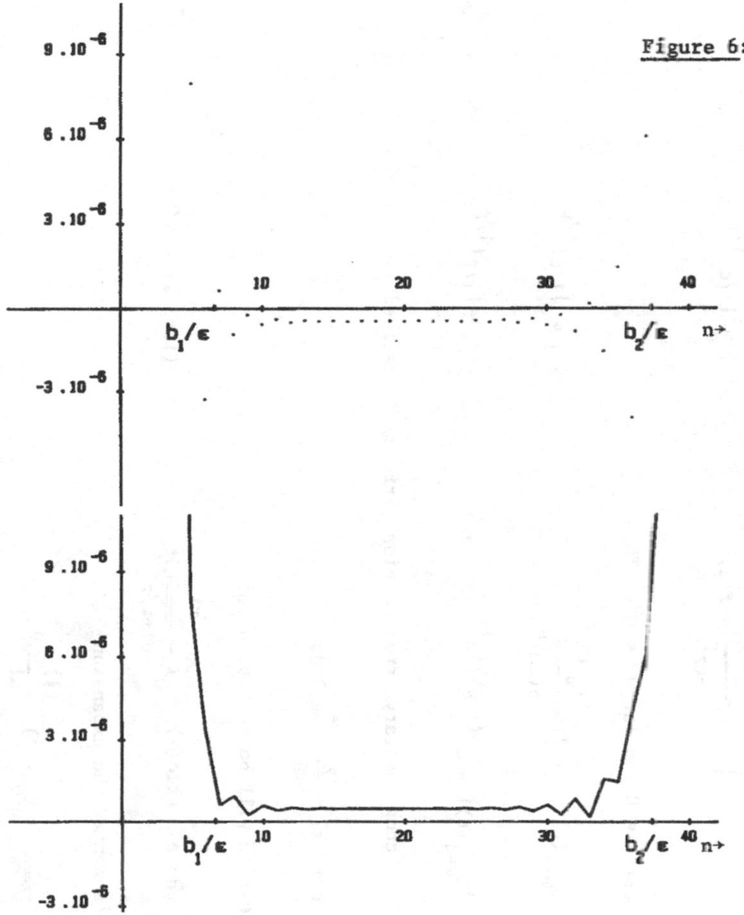

Figure 6: The incomplete Borel transform $F_a(\varepsilon) = \int_0^{a/\varepsilon} \frac{e^{-t}}{1+\varepsilon t} dt$ approximated by the sums $S_n(\varepsilon) = \sum_{k=0}^{n} (-1)^k k!\varepsilon^k$ with $\varepsilon = 0.05$ and $a = 0.7$. The complete Borel transform $F(x) = \int_0^\infty \frac{e^{-t}}{1+xt} dt$ and the incomplete Borel transform $F_a(x)$ share the asymptotic expansion $\sum_{k=0} (-1)^k k! x^k$, but they are differently approximated by its partial sums. In the case of the complete Borel transform they oscillate around the exact value: the remainders are nearly equal to the first neglected term of the expansion. In the case of the incomplete Borel transform they stick at a distance to the exact value which is approximately equal to the difference between $F_a(\varepsilon)$ and $F(\varepsilon)$. This level is attained when the terms are less than or of the order of this difference. Because of the estimations $F_a(\varepsilon) - F(\varepsilon) = £e^{-a/\varepsilon}$ and $|T_n(\varepsilon)| = e^{-[n\varepsilon(1-\log n\varepsilon + \emptyset)]/\varepsilon}$ the corresponding indices roughly lie between b_1/ε and b_2/ε (b_1 resp. b_2 are the lesser resp. the higher solution of $u(1-\log u) = a$). The figure above shows the remainders $R_n^a(\varepsilon) = F_a(\varepsilon) - S_n(\varepsilon)$ as a function of n, the figure below shows their absolute value $|R_n^a(\varepsilon)|$, joined by straight lines.

3. Generalizations.

The first generalization (Theorem 2.22) concerns the behaviour of the coefficients $(c_n)_{n\in\mathbb{N}}$. We allow regularily spaced zeros, that is to say the nonzero terms are placed at distance $p \geq 0$, and furthermore that c_{n+p}/c_n tends to an arbitrary limit. The theorem generalizes the formulas of Theorems 2.14, 2.15 and 2.16. The proofs of these theorems are easily generalized to a proof of Theorem 2.22.

Theorem 2.22: Let $(c_n)_{n\in\mathbb{N}}$ be a standard sequence and $p,q \in \mathbb{N}$. Suppose there exist a sequence $(pk+q)_{k\in\mathbb{N}}$ such that $\lim_{k\to\infty} c_{p(k+1)}/c_{pk+q} = c$, where $c < 0$ and a number $\bar{n}$ such that $c_n = 0$ for all indices $n \geq \bar{n}$ not of the form pk+q. Let $\varphi(z)$, τ_n, ρ_n and $f(z)$, t_n, r_n and $F(z)$, T_n, R_n be as usual. Let $\omega \in \mathbb{N}$ be of the form pk+q. Then

(i)

$$\rho_{\omega-1}(z) = \begin{cases} (1+\emptyset)\dfrac{\tau_\omega(z)}{1-c(z/\omega)^p} & |z/\omega| \lnsim |1/c|^{1/p} \\ (1+\emptyset)\dfrac{\tau_\omega(z)}{1-c(z/\omega)^p} + \varphi(z) & |z/\omega| \gnsim |1/c|^{1/p} \end{cases}$$

(ii)

$$r_{\omega-1}(z) = \begin{cases} (1+\emptyset)\dfrac{t_\omega(z)}{1-cz^p} & |z| \lnsim |1/c|^{1/p} \\ (1+\emptyset)\dfrac{t_\omega(z)}{1-cz^p} + f(z) & |z| \gnsim |1/c|^{1/p} \end{cases}$$

(iii) Suppose $\varepsilon \simeq 0$ and $|\arg \varepsilon| \lnsim \pi/p$. Then

$$R_{\omega-1}(\varepsilon) = (1+\emptyset)\frac{T_\omega(\varepsilon)}{1-c(\varepsilon\omega)^p} \qquad |\omega\varepsilon| \not\simeq |1/c|^{1/p}$$

$$|R_{\omega-1}(\varepsilon)| = e^{-(1+\emptyset)/|c^{1/p}\varepsilon|} \qquad |\omega\varepsilon| \simeq |1/c|^{1/p}$$

Example. The "complementary errorfunction" erfc(z) is defined by

$$\operatorname{erfc}(z) = \frac{2}{\sqrt{\pi}} \int_z^\infty e^{-s^2} ds$$

This expression will be rewritten to

$$\sqrt{\pi}\, z\, e^{z^2} \operatorname{erfc}(z) = \int_0^\infty \frac{e^{-t}}{\sqrt{1+t/z^2}} dt \qquad (|\arg z| < \pi/2)$$

By Watson's lemma, the expansion

$$(1) \qquad \frac{1}{\sqrt{1+w}} = \sum_{n=0}^\infty (-1)^n \frac{(\frac{1}{2})_n}{n!} w^n$$

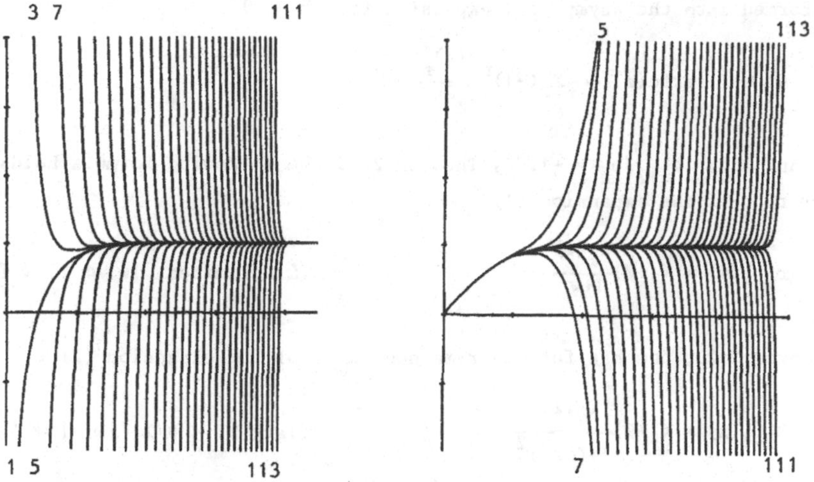

Figure 7: Divergent and convergent expansions of the errorfunction. The "errorfunction" erf(x) is defined by

$$\mathrm{erf}(x) = \frac{2}{\sqrt{\pi}} \int_0^x e^{-s^2} ds$$

Its complement (with respect to 1) is the "complementary errorfunction" $\mathrm{erfc}(x) = \frac{2}{\sqrt{\pi}} \int_x^\infty e^{-s^2} ds$, treated in the text. As follows from the text, the errorfunction has a divergent expansion at infinity with partial sums

$$S_{2n+1}(x) = 1 - \frac{e^{-x^2}}{\sqrt{\pi}} \sum_{k=0}^{n} (-1)^k \frac{(\frac{1}{2})_k}{x^{2k+1}}$$

The figure on the left shows the graphs of these sums for 2n+1 = 1 to 2n+1 = 113. As a function of n, the graphs of the S_n "retreat" towards infinity, it can be shown by the methods of Chapter III, section 4, that the horizontal distance between S_{n+4} and S_n is of order $1/\sqrt{n}$. The figure on the right shows the graphs of the polynomials $P_n(x)$ of the convergent Taylor expansion of erf(x),

$$P_{2n+1}(x) = \frac{2}{\sqrt{\pi}} \sum_{k=0}^{n} (-1)^n \frac{x^{2n+1}}{n!(2n+1)}$$

for 2n+1 = 5 to 2n+1 = 113. The graphs "progress" towards infinity. Again the shift from P_n to P_{n+4} is of order $1/\sqrt{n}$, but it can be shown [12], that it is somewhat larger than the shift from S_n to S_{n+4}.

is transformed into the asymptotic expansion (for $z \to \infty$)

(2) $$\sqrt{\pi}\, z\, e^{z^2} \operatorname{erfc}(z) \sim \sum_{n=0} (-1)^n \frac{(\frac{1}{2})_n}{z^{2n}}$$

Here p=2 and $\lim_{n \to \infty} c_{n+2}/c_n = -1$. By Theorem 2.15 the following formula holds for the remainder $r_{\omega-1}$ of the expansion (1)

(3) $$r_{\omega-1}(w) = (1+\emptyset)\frac{t_\omega(w)}{1+w} \qquad (\omega \text{ unlimited, } |w| \not\simeq 1,\ w \notin]-\infty,-1])$$

So by Theorem 2.20, one has for the remainder $R_{\omega-1}$ of the expansion (2)

(4) $$R_{\omega-1}(z) = (1+\emptyset)\frac{T_\omega(z)}{1+(\omega/z)^2} \qquad (|\arg z| \lessapprox \pi/2,\ |\omega/z| \not\simeq 1,\ \omega \text{ even})$$

In fact, sharper results can be proved. By Proposition 1.6 formula (3) still holds for $|w| \simeq 1$, $w \not\simeq -1$, and by Theorem 3.2 formula (4) also holds for $|\omega/z| \simeq 1$.

A second generalization concerns integrals of type

$$F^r(z) = \int_0^\infty e^{-t} t^r f(zt)\,dt$$

$$F_z{}^r(z) = \int_0^{a/z} e^{-t} t^r f(zt)\,dt \qquad (r \text{ limited, } r \gtrapprox -1)$$

The formulas of Theorems 2.16, 2.18 and 2.19 remain unchanged for these integrals (see for instance Proposition 1.7 . Integrals of this form arise in relation to confluent hypergeometric functions and Besselfunctions.

We now generalize Theorem 2.15 for sequences of coefficients with mildly irregular behaviour.

<u>Definition 2.23</u>: Let $(a_n)_{n\in\mathbb{N}}$ be a sequence and $n \in \mathbb{N}$. The first nonzero term of the sequence of index larger than n will be written $c_{n+p(n)}$.

<u>Proposition 2.24</u>: Let $f(z) = \sum_{n=0}^{\infty} c_n z^n$ be standard. Let $R = \liminf_{n \to \infty} |c_n/c_{n+p(n)}|^{1/p(n)}$. Let ω be unlimited such that $c_\omega \neq 0$ and $A_\omega(z)$ be the approximation factor of index ω. Suppose $0 < R < \infty$. Then for all $|z| \lessapprox R$

$$A_\omega(z) \simeq \sum_{n=0}^{\infty} d_{n,\omega} z^n$$

where $\sum_{n=0}^{\infty} d_{n,\omega} z^n$ is a standard series with convergence radius $\geqq R$.

<u>Proof</u>: For $|z| < R$ the series $\sum_{n=0}^{\infty} c_n z^n$ is convergent. So we have the formula

$$A_\omega(z) = \sum_{n=0}^{\infty} \frac{c_{\omega+n}}{c_n} z^n$$

Now $c_{\omega+n}/c_\omega$ is limited for all limited n. So we may define

$$d_{n,\omega} = {}^o(c_{\omega+n}/c_\omega)$$

Let $(d_{n,\omega})_{n\in\mathbb{N}}$ be the standardized of $(d_{n,\omega})_{st\ n\in\mathbb{N}}$. Then $\sum_{n=0}^{\infty} d_{n,\omega} z^n$ is a standard series with convergence radius $\geq R$. The approximation $A_\omega(z) = \sum_{n=0}^{\infty} d_{n,\omega} z^n$ follows from the lemma of dominated approximation: indeed, if $|z| \lneq R$, then $\frac{c_{\omega+n}}{c_\omega} z^n \simeq d_{n,\omega} z^n$ for limited n and $\left|\frac{c_{\omega+n}}{c_\omega} z^n\right| \leq \left|\frac{{}^o|z|+R}{2R}\right|^n$ for all n.

The last generalization concerns divergent series with coefficients with growth type $(n!)^{1/k}$, where $k > 0$. Such series are called Gevrey. It can be shown [20], [52], [59] that formal series solutions of rational differential equations

$$\frac{dY}{dX} = \frac{Q(X,Y)}{P(X,Y)} \qquad \text{P, Q polynomials}$$

are usually of this type. As shown by Ramis [58], these formal series may be written as a finite sum of series $\sum_{n=0}^{\infty} c_n (n!)^{1/k}/X^n$ with $c_{n+1}/c_n \to c$.

Consider a series $f(z) = \sum_{n=0}^{\infty} \frac{c_n (n!)^{1/k}}{\Gamma(1+n/k)} z^n$, which is convergent with convergence radius, say, R ($0 < R < \infty$). Its incomplete Leroy-transform $l_{k,a}(f)(z)$ is defined by

$$l_{k,a}(f)(z) = \frac{k}{z^k} \int_0^a e^{-(t/z)^k} t^{k-1} f(t)dt \qquad (a < R)$$

The incomplete Leroy-transform satisfies formally the expansion

$$l_{k,a}(f)(z) \sim \sum_{n=0} c_n (n!)^{1/k} z^n$$

Next theorem indicates how real incomplete Leroy-transforms are approximated by the partial sums of this series.

Theorem 2.25: Let $\sum_{n=0} c_n (n!)^{1/k} x^n$ be a standard formal series, where $\lim_{n\to\infty} c_{n+1}/c_n = -1$ and $k > 0$. Let $\varepsilon > 0$ be infinitesimal. Let $0 \lneq a \lneq (1/k)^{1/k}$ and put $f(x) = \sum_{n=0}^{\infty} \frac{c_n (n!)^{1/k} x^n}{\Gamma(1+n/k)}$, $T_n(\varepsilon) = c_n (n!)^{1/k} \varepsilon^n$ and $R_{n-1}{}^a(\varepsilon) = l_{k,a}(f)(\varepsilon) - \sum_{m=0}^{n-1} T_m(\varepsilon)$. Let a_1 ($\lneq 1$) be the lesser solution of $u(1-\log u) = ka^k$ and a_2 ($\gneq 1$) be the greater solution of this equation. Then

$$R_{n-1}{}^a(\varepsilon) = (1+\emptyset)\frac{T_n(\varepsilon)}{1+\varepsilon n^{1/k}} \qquad n\,\varepsilon^k \lneq a_1, \text{ or } n\,\varepsilon^k \gneq a_2$$

$$|R_{n-1}{}^a(\varepsilon)| = e^{-\left(\frac{a+\emptyset}{\varepsilon}\right)^k} \qquad a_1 \lessapprox n\,\varepsilon^k \lessapprox a_2$$

<u>Proof</u>: Put $d_n = \dfrac{c_n(n!)^{1/k}}{\Gamma(1+n/k)}$, $t_n(x) = d_n x^n$ and $r_n(x) = f(x) - \sum_{m=0}^{n} d_m x^m$. Let $\omega \in \mathbb{N}$ be unlimited. It follows from Stirling's formula that $d_{\omega+1}/d_\omega \simeq -k^{1/k}$. Hence the convergence radius of $\sum_{n=0}^{\infty} d_n x^n$ is $(1/k)^{1/k}$. By Theorem 2.15 we have for $x \lnapprox (1/k)^{1/k}$

$$r_{\omega-1}(x) = (1+\oslash)\frac{t_\omega(x)}{1+k^{1/k}x}$$

Let $ka^k \lnapprox \bar{c} \lnapprox 1$. Let $c = (\bar{c}/k)^{1/k}$. Then $a \lnapprox c \lnapprox (1/k).^{1/k}$. As in the proof of Theorem 2.20 we write

$$U_1 = \frac{k}{\varepsilon^k}\int_0^c e^{-(t/\varepsilon)^k} r_{n-1}(t)t^k dt$$

$$U_2 = \sum_{m=0}^{n-1} d_m \frac{k}{\varepsilon^k}\int_c^\infty e^{-(t/\varepsilon)^k} t^{m+k} dt$$

$$U_3 = 1_{k,a}(f)(\varepsilon) - 1_{k,c}(f)(\varepsilon)$$

Hence $R_{n-1}{}^a(\varepsilon) = U_1+U_2+U_3$. We have $U_3 = (e^{-\frac{a+\oslash}{\varepsilon}})^k$. For limited n the term U_2 is also exponentially small, and one can show that $U_1 = (1+\oslash)T_n(\varepsilon)$ along the lines of the proof of Watson's lemma. Hence $R_{n-1}{}^a(\varepsilon) = (1+\oslash)T_n(\varepsilon) = (1+\oslash)\dfrac{T_n(\varepsilon)}{1+\varepsilon n^{1/k}}$. For unlimited n we have the estimation, obtained from Stirling's formula

$$T_n(\varepsilon) = e^{-n\,\varepsilon^k(1-\log(n\,\varepsilon^k)+\oslash)/k\varepsilon^k}$$

Now the following estimations can be shown in the same manner as in the proof of Theorem 2.20, taking into account the above estimations for U_3 and $T_n(\varepsilon)$

	U_1	U_2	$R_{n-1}(\varepsilon) = U_1+U_2+U_3$
$n\,\varepsilon^k \lnapprox a_1$	$(1+\oslash)\dfrac{T_n(\varepsilon)}{1+\varepsilon n^{1/k}}$	$e^{-(c/\varepsilon)^k}(f(c)+\oslash)$	$(1+\oslash)\dfrac{T_n(\varepsilon)}{1+\varepsilon n^{1/k}}$
$a_1 \lessapprox n\,\varepsilon^k \lessapprox a_2$	$£e^{-(\frac{a+\oslash}{\varepsilon})^k}$	$£e^{-(\frac{a+\oslash}{\varepsilon})^k}$	$\Theta e^{-(\frac{a+\oslash}{\varepsilon})^k}$, $\lvert\Theta\rvert = 1$
$n\,\varepsilon^k \gnapprox a_2$	$£e^{-(\frac{a+\oslash}{\varepsilon})^k}$	$(1+\oslash)\dfrac{T_n(\varepsilon)}{1+\varepsilon n^{1/k}}$	$(1+\oslash)\dfrac{T_n(\varepsilon)}{1+\varepsilon n^{1/k}}$

The estimations of the last column prove the theorem.

In this chapter we carefully avoided some "critical regions", which, however, were rather small. For everywhere convergent expansions we had to avoid the points which made the terms nearly maximal, for expansions with convergence radius we had to stay away from the vicinity of the convergence circle, and for divergent expansions we had to avoid the indices near the index of the smallest term. In the next chapter we will enter these regions, but it is at the price of stronger conditions and much harder labor.

CHAPTER III. ASYMPTOTIC EXPRESSIONS FOR THE REMAINDERS ASSOCIATED TO EXPANSIONS OF TYPE $\sum_{n=0}^{\infty} c_n \frac{z^n}{n!}$, $\sum_{n=0}^{\infty} c_n z^n$ AND $\sum_{n=0} c_n n! z^n$: CRITICAL REGIONS, UNIFORM BEHAVIOUR.

Under additional conditions, the validity of the asymptotic formulas of the remainder derived in the previous chapter, may be extended to the "critical regions", where they could not be established by elementary methods. The "critical regions" are: for entire functions $\sum_{n=0}^{\infty} c_n \frac{z^n}{n!}$ those z where $c_n \frac{z^n}{n!}$ takes its largest values, for meromorph functions $\sum_{n=0}^{\infty} c_n z^n$ the halo of the convergence circle, and for divergent series $\sum_{n=0} c_n n! \varepsilon^n$, with $\varepsilon \simeq 0$ fixed, the indices close to the smallest term.

In Section 1 we show that if the validity of the asymptotic formulas can be extended, we can answer the question of (nearly) optimal approximation by divergent series. Two kinds of additional conditions leading to such an extension are examined in Sections 2 and 3. In the last section a further application is presented: the remainders of high order of the expansions under consideration exhibit a strong uniform behaviour as a function of the distance to the origine.

Remark: In this chapter we maintain the notations of chapter II (see 2.1 and 2.10).

1. Summation to the nearly minimal term of divergent series of type $\sum_{n=0} c_n n! \varepsilon^n$.

Definition 3.1 (i) Let $(U_n)_{n\in\mathbb{N}}$ be a sequence of real or complex numbers. A term U_m will be called nearly minimal if $|U_m/U_n| \lesssim 1$ for all $n \in \mathbb{N}$. Of course only nonzero terms are taken into consideration.

(ii) Let f be a number, $(T_n)_{n\in\mathbb{N}}$ be a sequence of terms and $(R_n)_{n\in\mathbb{N}}$, defined by $R_n = f - \sum_{k=0}^{n} T_k$, be a sequence of remainders. The approximation $f = \sum_{k=0}^{n} T_k + R_n$ is said to satisfy the property of summation to the nearly minimal term if R_{n-1} is nearly minimal if and only if T_n is nearly minimal. Furthermore, the least index such that $|T_k| \leq |T_n|$ for all $n \in \mathbb{N}$ will be written K and the least index such that $|R_k| \leq |R_n|$ for all $n \in \mathbb{N}$ will be written L.

Let $f(z) = \sum_{n=0}^{\infty} c_n z^n$, where $\lim_{n\to\infty} c_{n+1}/c_n = -1$, and let $F(z) = \int_0^\infty e^{-t} f(zt)dt$ be its Borel transform. In Chapter II it was shown (Theorem 2.15) that for unlimited ω the approximation factor $A_\omega^f(z) = r_{\omega-1}(z)/t_\omega(z)$ is nearly equal to $\frac{1}{1+z}$, outside the halo of the circle of convergence. This was sufficient to show that for the divergent expansion $F(z) \sim \sum_{n=0} c_n n! z^n$, and for $z = \varepsilon \simeq 0$, the index L is of the form $(1+\emptyset)/|\varepsilon|$. Also K is of this form. However, K and L still may lay rather far apart. Next theorem shows that if the approximation $A_\omega^f(z) \simeq \frac{1}{1+z}$ also holds within the halo of the convergence circle, a sharper result is possible: the indices of the nearly minimal terms and the nearly minimal remainders coincide.

<u>Theorem 3.2</u>: Let f(z) be the analytic continuation of a series $\sum_{n=0}^{\infty} c_n z^n$, where $\lim_{n\to\infty} c_{n+1}/c_n = -1$. Let $\varepsilon \simeq 0$ be such that $|\arg \varepsilon| \lneq \pi$, and such that f is of S-exponential order in this direction. Suppose $A_\omega(z) \simeq \frac{1}{1+z}$ for all unlimited $\omega \in \mathbb{N}$ and limited z in the direction of $\arg(\varepsilon)$. Let $F(\varepsilon)$, $T_n(\varepsilon)$ and $R_{n-1}(\varepsilon)$ be as usual. Then

(i) for all $n \in \mathbb{N}$ one has $R_{n-1}(\varepsilon) = (1+\emptyset)\frac{T_n(\varepsilon)}{1+n\varepsilon}$

(ii) A remainder $R_{k-1}(\varepsilon)$ is nearly minimal iff the term $T_k(\varepsilon)$ is nearly minimal.

<u>Proof</u>: (i) We only need to consider the case $|n\varepsilon| \simeq 1$. It follows from Theorem 2.15 that for limited $a \gneq 1$

$$r_{n-1}(z) = \begin{cases} (1+\emptyset)t_n(z)/(1+z) & |z| \le |a| \\ (1+\emptyset)t_n(z)/(1+z)+f(z) & |z| \geqq |a| \end{cases}$$

Hence

$$\begin{aligned} R_{n-1}(\varepsilon) &= \int_0^\infty e^{-t} r_{n-1}(\varepsilon t)dt \\ &= c_n\varepsilon^n \int_0^\infty e^{-t}(1+\alpha(t))\frac{t^n}{1+\varepsilon t}dt + \int_{|a/\varepsilon|}^\infty e^{-t}f(\varepsilon t)dt \qquad (\alpha(t) \simeq 0) \\ &= (1+\emptyset)\frac{T_n(\varepsilon)}{1+n\varepsilon} + \Theta c^{-|a+\emptyset/\varepsilon|} \qquad (|\Theta| = 1) \\ &= (1+\emptyset)\frac{T_n(\varepsilon)}{1+n\varepsilon} \end{aligned}$$

(ii) As a consequence of Theorem 2.16, both the indices of the nearly minimal terms and remainders are of the form $(1+\emptyset)/|\varepsilon|$. Now if k and n are of this form, we have by (i)

$$\frac{R_n(\varepsilon)}{R_k(\varepsilon)} = (1+\emptyset)\frac{T_{n+1}(\varepsilon)(1+(k+1)\varepsilon)}{T_{k+1}(\varepsilon)(1+(n+1)\varepsilon)} = (1+\emptyset)\frac{T_n(\varepsilon)}{T_k(\varepsilon)}$$

Hence $R_k(\varepsilon)$ is nearly minimal iff $T_k(\varepsilon)$ is nearly minimal.

<u>Comments</u>: 1) In Chapter II we proved that both $|T_\omega(\varepsilon)|$ and $|R_\omega(\varepsilon)|$ have the approximate value $e^{-|\omega\varepsilon|(1-\log|\omega\varepsilon|+\emptyset)/|\varepsilon|}$. So it is to be expected that for small ε the remainders are already very small for indices considerably less than the index (about $|1/\varepsilon|$) of the smallest term. This is illustrated by the third column of

Fig. 3 of Chapter II. For many practical purposes the "rough" results of Chapter II suffice. However, the refinements of this chapter may have practical implications for relative big ε's, say of order 1/10. They also are of theoretical interest.

2) In order to be able to conclude that $A_n^F(\varepsilon) \simeq \frac{1}{1+n\varepsilon}$ for $|n\varepsilon| \simeq 1$ we need to know that $A_\omega^f(z) \simeq \frac{1}{1+z}$ for z within the halo of the convergence circle. What is more, one only needs to know it there. Indeed, by the concentration lemma 5.6 there exists $\beta \simeq 0$ such that

$$\int_0^\infty e^{-t} r_{n-1}(\varepsilon t)dt = c_n \varepsilon^n \int_0^\infty e^{-t} t^n A_n^f(\varepsilon t)dt$$

$$= c_n \varepsilon^n \int_{n(1-\beta)}^{n(1+\beta)} e^{-t} t^n A_n^f(\varepsilon t)dt + \mu \qquad (\mu \in \frac{1}{n}\text{-M})$$

On $[n(1-\beta), n(1+\beta)]$ the function $A_n^f(\varepsilon t) \simeq \frac{1}{1+\varepsilon t}$ is nearly constant $(\simeq \frac{1}{1+n\varepsilon})$. See also the proof of Theorem 1.4.

3) Up to now we carefully left out the distance arg ε = π. Indeed, it is not possible to take the Borel transform of f in this direction for f has a singularity in -1. In the example below this singularity is a simple pole and we show that the property of "summation to the smallest term" holds for the Cauchy-principal value of the Borel transform.

<u>Example: asymptotic expansion and best approximation of the Cauchy principal value of a Borel integral.</u>

<u>Proposition 3.3</u>: Let $\varepsilon > 0$. The integral

$$F(\varepsilon) = ⨍_0^\infty \frac{e^{-t}}{1-\varepsilon t}dt$$

has the asymptotic expansion $\sum_{n=0}^\infty n!\varepsilon^n$. If $\varepsilon \simeq 0$, the remainder $R_{[1/\varepsilon]-1}$ or $R_{[1/\varepsilon]}$ is minimal. In absolute value, the minimal remainder is less than $(\frac{1}{2}+\eta)\sqrt{\frac{\pi}{2\varepsilon}}e^{-1/\varepsilon}$, where $\eta \simeq 0$.

<u>Comments</u>: 1) All terms of the expansion have the same sign. This implies that the way the number F(ε) is approached by the partial sums is entirely different from the alternating case (here $\varepsilon < 0$) or the complex case. In the complex case the partial sums are turning around the exact value, in the alternating case the angular velocity is π, and they jump backward and forward over the exact value. In both cases there often is no clearly defined best remainder: under the conditions of Theorem 3.2 all remainders which have the index of a nearly minimal term are nearly minimal themselves. But here the partial sums are constantly growing, so there is a clearly

defined best remainder: best approximation is obtained when the remainder changes sign. Notice that in the example this corresponds to the index (±1) of the smallest term.

2) The problem was already solved by Stieltjes in 1886 -using infinitesimals. The proof below is not very far from that by Stieltjes.

Notations: Put

$$F(\alpha,\varepsilon) = \int_0^{(1-\alpha)/\varepsilon} \frac{e^{-t}}{1-\varepsilon t}dt + \int_{(1+\alpha)/\varepsilon}^{\infty} \frac{e^{-t}}{1-\varepsilon t}dt$$

Notice that $F(\varepsilon) = \lim\limits_{\alpha\downarrow 0} F(\alpha,\varepsilon)$. The number ε and the parameter α will always be assumed positive and infinitesimal. From the identity $\frac{1}{1-t} = \sum_{k=0}^{n-1} t^k + \frac{t^n}{1-t}$ it follows that

$$F(\alpha,\varepsilon) = \sum_{k=0}^{n-1} k!\varepsilon^k + \varepsilon^n \int_0^{(1-\alpha)/\varepsilon} e^{-t} \frac{t^n}{1-\varepsilon t}dt + \varepsilon^n \int_{(1+\alpha)/\varepsilon}^{\infty} e^{-t} \frac{t^n}{1-\varepsilon t}dt - \sum_{k=0}^{n-1}\varepsilon^k \int_{(1-\alpha)/\varepsilon}^{(1+\alpha)/\varepsilon} e^{-t}t^k dt$$

Put

$$I_1 = \varepsilon^n \int_0^{(1-\alpha)/\varepsilon} e^{-t} \frac{t^n}{1-\varepsilon t}dt$$

$$I_2 = \varepsilon^n \int_{(1+\alpha)/\varepsilon}^{\infty} e^{-t} \frac{t^n}{1-\varepsilon t}dt$$

$$I_3 = \sum_{k=0}^{n-1} \varepsilon^k \int_{(1-\alpha)/\varepsilon}^{(1+\alpha)/\varepsilon} e^{-t}t^k dt$$

$$R_{n-1}(\alpha,\varepsilon) = I_1 + I_2 + I_3$$

For the proof of proposition 3.3 we need some lemma's.

Lemma 1: If $n = 1/\varepsilon + p$, where p is limited, then there exists $\beta \simeq 0$ such that

$$I_1 + I_2 = -2\,\frac{e^{-1/\varepsilon}}{\sqrt{\varepsilon}}\left(\int_{\alpha/\sqrt{\varepsilon}}^{\beta/\sqrt{\varepsilon}} e^{-u^2/2}(p+\oslash+u^2/3(1+\oslash))du+\mu\right)$$

where $\mu \in \varepsilon$-M.

Proof: First we put the singularity in 0. Substituting $t = (1-s)/\varepsilon$ we obtain

$$I_1 = \frac{e^{-1/\varepsilon}}{\varepsilon}\int_\alpha^1 e^{(s+\log(1-s))/\varepsilon}\,\frac{(1-s)^p}{s}\,ds$$

By the concentration lemma 5.6 there exists $\beta \simeq 0$ such that $\beta/\sqrt{\varepsilon}$ is unlimited and

$$I_1 = \frac{e^{-1/\varepsilon}}{\varepsilon}[\int_\alpha^\beta e^{(s+\log(1-s))/\varepsilon}\frac{(1-s)^p}{s}ds+\mu_1]$$

with $\mu_1 \in \varepsilon$-M. Now the approximations $s+\log(1-s) = -\frac{s^2}{2}+\frac{s^3}{3}(1+\emptyset)$ and $(1-s)^p = 1-ps(1+\emptyset)$ hold for all s on the interval of integration. Put $s = u\sqrt{\varepsilon}$. Then

$$I_1 = \frac{e^{-1/\varepsilon}}{\varepsilon}\int_{\alpha/\sqrt{\varepsilon}}^{\beta/\sqrt{\varepsilon}} \frac{e^{-u^2/2}(1-\frac{u^3}{3}\sqrt{\varepsilon}(1+\emptyset))(1-pu\sqrt{\varepsilon}(1+\emptyset))}{u}du+\mu_1$$

$$= \frac{e^{-1/\varepsilon}}{\varepsilon}\int_{\alpha/\sqrt{\varepsilon}}^{\beta/\sqrt{\varepsilon}}\frac{e^{-u^2/2}}{u}du - \frac{e^{-1/\varepsilon}}{\sqrt{\varepsilon}}[\int_{\alpha/\sqrt{\varepsilon}}^{\beta/\sqrt{\varepsilon}} e^{-u^2/2}(p+\emptyset+u^2/3(1+\emptyset))du+\mu_2]$$

with $\mu_2 \in \varepsilon$-M. In the same manner we obtain

$$I_2 = -\frac{e^{-1/\varepsilon}}{\varepsilon}\int_{\alpha/\sqrt{\varepsilon}}^{\beta/\sqrt{\varepsilon}}\frac{e^{-u^2/2}}{u}du - \frac{e^{-1/\varepsilon}}{\sqrt{\varepsilon}}[\int_{\alpha/\sqrt{\varepsilon}}^{\beta/\sqrt{\varepsilon}} e^{-u^2/2}(p+\emptyset+u^2/3(1+\emptyset))du+\mu_3]$$

with $\mu_3 \in \varepsilon$-M. By adding I_1 and I_2 the "dangerous" term $\frac{e^{-1/\varepsilon}}{\sqrt{\varepsilon}}\int_{\alpha/\sqrt{\varepsilon}}^{\beta/\sqrt{\varepsilon}}\frac{e^{-u^2/2}}{u}du$ disappears and one gets

$$I_1+I_2 = -2\frac{e^{-1/\varepsilon}}{\sqrt{\varepsilon}}[\int_{\alpha/\sqrt{\varepsilon}}^{\beta/\sqrt{\varepsilon}} e^{-u^2/2}(p+\emptyset+u^2/3(1+\emptyset))du+\mu] \qquad (\mu \in \varepsilon\text{-M})$$

Lemma 2: If $n = 1/\varepsilon+p$, where p is limited, andif $\alpha/\sqrt{\varepsilon} \simeq 0$, then

$$I_1+I_2 = -\sqrt{\frac{2\pi}{\varepsilon}}e^{-1/\varepsilon}(p+\frac{1}{3}+\emptyset)$$

Proof: We estimate the integral with the aid of the lemma of dominated approximation Put

$$I(u) = \begin{cases} 0 & u < \alpha/\sqrt{\varepsilon} \\ e^{-u^2/2}(p+\emptyset+u^2/3.(1+\emptyset)) & \alpha/\sqrt{\varepsilon} \leqq u \leqq \beta/\sqrt{\varepsilon} \\ 0 & u > \beta/\sqrt{\varepsilon} \end{cases}$$

Then $I(u) \simeq e^{-u/2}(p+u^2/3)$ for all positive appreciable u and $|I(u)| \leqq e^{-u^2/2}(|p|+1+u^2)$ for all $u \geqq 0$. Then by the lemma of dominated approximation

$$\int_0^\infty I(u)du \simeq \int_0^\infty e^{-u^2/2}(p+u^2/3)du = \sqrt{\frac{\pi}{2}}(p+1/3)$$

Hence $I_1+I_2 = -\sqrt{\frac{2\pi}{\varepsilon}}\, e^{-1/\varepsilon}(p+\frac{1}{3}+\emptyset)$.

Lemma 3: If $n = 1/\varepsilon+p$, where p is limited, and if $\alpha/\varepsilon\sqrt{\varepsilon} \simeq 0$, then

$$I_3 = \emptyset.\frac{e^{-1/\varepsilon}}{\sqrt{\varepsilon}}$$

Proof: One has

$$I_3 = \sum_{k=0}^{1/\varepsilon+p-1} \varepsilon^k \int_{(1-\alpha)/\varepsilon}^{(1+\alpha)/\varepsilon} e^{-t}t^k dt$$

$$\leq (1/\varepsilon+p).\frac{2\alpha}{\varepsilon}e^{-(1-\alpha)/\varepsilon}(1+\alpha)^{1/\varepsilon+p-1}$$

$$\leq (2+\emptyset)\frac{\alpha}{\varepsilon^2}e^{-1/\varepsilon+2\alpha/\varepsilon}$$

Hence $I_3 = \emptyset \frac{e^{-1/\varepsilon}}{\sqrt{\varepsilon}}$ if $\alpha = \emptyset.\varepsilon\sqrt{\varepsilon}$

Corollary 4: If $n = 1/\varepsilon+p$, where p is limited, and $\alpha/\varepsilon\sqrt{\varepsilon} \simeq 0$, then

$$R_n(\alpha,\varepsilon) = -\sqrt{\frac{2\pi}{\varepsilon}}\, e^{-1/\varepsilon}(p+\frac{1}{3}+\emptyset)$$

Lemma 5: If $\alpha/\sqrt{\varepsilon} \simeq 0$, then $F(\varepsilon)-F(\alpha,\varepsilon) = \emptyset \frac{e^{-1/\varepsilon}}{\sqrt{\varepsilon}}$

Proof: After some algebra one finds

$$F(\alpha,\varepsilon) = \frac{e^{-1/\varepsilon}}{\sqrt{\varepsilon}}\left(\int_\alpha^{1/\varepsilon} \frac{e^s-e^{-s}}{s}ds + \int_{1/\varepsilon}^{\infty} \frac{e^{-s}}{s}ds\right)$$

Hence for $\alpha/\sqrt{\varepsilon} \simeq 0$

$$F(\varepsilon)-F(\alpha,\varepsilon) = \frac{e^{-1/\varepsilon}}{\sqrt{\varepsilon}} \int_0^\alpha \frac{e^s-e^{-s}}{s}ds$$

$$= \frac{e^{-1/\varepsilon}}{\sqrt{\varepsilon}} \int_0^\alpha \frac{2s(1+\emptyset)}{s}ds$$

$$= (2+\emptyset)\frac{\alpha}{\varepsilon}\, e^{-1/\varepsilon} = \emptyset.\frac{e^{-1/\varepsilon}}{\sqrt{\varepsilon}}$$

Proof of proposition 3.3: We first prove that $R_{[1/\varepsilon]-1}$ or $R_{[1/\varepsilon]}$ is minimal and approximate their value. Let α be such that $\alpha/\varepsilon\sqrt{\varepsilon} \simeq 0$. Then

$$R_{1/\varepsilon+p-1}(\varepsilon) = F(\varepsilon)-F(\alpha,\varepsilon)+R_{1/\varepsilon+p-1}(\alpha,\varepsilon)$$

$$= \emptyset.\frac{e^{-1/\varepsilon}}{\sqrt{\varepsilon}} + (p+\frac{1}{3}+\emptyset)\sqrt{2\pi}\,\frac{e^{-1/\varepsilon}}{\sqrt{\varepsilon}}$$

$$= (p+\frac{1}{3}+\emptyset)\sqrt{\frac{2\pi}{\varepsilon}}\, e^{-1/\varepsilon}$$

Hence, depending on the value of $1/\varepsilon-[1/\varepsilon]$, the remainder $R_{[1/\varepsilon]-1}$ or $R_{[1/\varepsilon]}$ is minimal. In absolute value, it does not exceed $(\frac{1}{2}+\emptyset)\sqrt{\frac{2\pi}{\varepsilon}}\, e^{-1/\varepsilon}$. Secondly we prove that $\sum_{k=0} k!\varepsilon^k$ is the asymptotic expansion of $F(\varepsilon)$ for $\varepsilon \downarrow 0$. Let $\varepsilon \simeq 0$, $\varepsilon > 0$. By the above estimation

$$F(\varepsilon) = \sum_{k=0}^{[1/\varepsilon]} k!\varepsilon^k + e^{-(1+\emptyset)/\varepsilon}$$

Suppose that $a \lessapprox 1$. By lemma 5.12, we have for st n

$$\sum_{k=n}^{[a/\varepsilon]} k!\varepsilon^k = (1+\emptyset)n!\varepsilon^n$$

By the Fehrele principle (4.12) there exists $\gamma \simeq 0$ such that still $\sum_{k=n}^{(1-\gamma)/\varepsilon} k!\varepsilon^k = (1+\emptyset)n!\varepsilon^n$. Now it follows easily from lemma 5.13(iii) that

$$\sum_{k=(1-\gamma)/\varepsilon+1}^{[1/\varepsilon]} k!\varepsilon^k = e^{-(1+\emptyset)/\varepsilon}$$

Combining

$$\begin{aligned} F(\varepsilon) &= \sum_{k=0}^{n-1} k!\varepsilon^k + \sum_{k=n}^{[1/\varepsilon]} k!\varepsilon^k + e^{-(1+\emptyset)/\varepsilon} \\ &= \sum_{k=0}^{n-1} k!\varepsilon^k + (1+\emptyset)n!\varepsilon^k \end{aligned}$$

Hence it follows from proposition 2.6(iii) that $\sum_{k=0} k!\varepsilon^k$ is the asymptotic expansion of $F(\varepsilon)$ for $\varepsilon \downarrow 0$.

<u>Comments</u>: 1) For all n such that $n\varepsilon \lessapprox 1$ we have

$$R_{n-1}(\varepsilon) = \frac{T_n(\varepsilon)}{1-n\varepsilon}$$

The last part of the proof of proposition 3.3 is easily adapted to this case (use lemma 5.11 instead of 5.12).

2) The example illustrates that remainders of asymptotic expansions do not need to be of the order of the first neglected term. Indeed, it will be shown that there are indices n less than the index of best approximation such that $T_n(\varepsilon)/R_{n-1}(\varepsilon) \simeq 0$. It follows from Stirling's formula that for all limited p

$$T_{1/\varepsilon+p}(\varepsilon) = (1+\emptyset)\sqrt{\frac{2\pi}{\varepsilon}}\, e^{-1/\varepsilon}$$

and thus

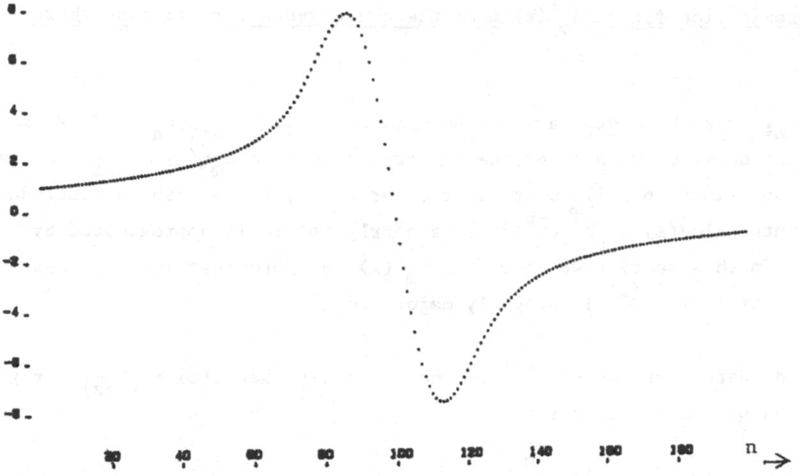

Figure 1: The approximation factor $A_n^F(x)$ associated to the asymptotic expansion $\sum_{n=0} n!x^{n+1}$ of the integral $xF(x) = \int_0^\infty \frac{e^{-t/x}}{1-t}dt$ for $x = 0.01$, and $0 \leq n \leq 200$. The approximation properties of divergent series with alternate terms and with terms of constant sign are different. In the first case typically the remainders are less than the first neglected term, the partial sums alternate about the exact value and there is no outspoken index of optimal approximation, the best indices are close to that of the smallest term. The second case is illustrated by the above drawing. In the beginning the remainders are larger than the first neglected term (theoretically $A_n^F(x) = \frac{1+\emptyset}{1-n\varepsilon}$ for $x \to 0$, $n\varepsilon \not\simeq 1$). For some indices, paradoxically just before optimal approximation is attained, the remainders are even not of order of the first neglected term. There is a sharply defined index of best approximation, situated just there, where the partial sums, constantly growing, overgrow the exact value. Here it equals the index of the smallest term (100). Then the approximation factor changes sign. For n well beyond $1/\varepsilon$ it again assumes the approximate value $\frac{1+\emptyset}{1-n\varepsilon}$.

$$R_{1/\varepsilon+p-1}(\varepsilon) = -(p+\frac{1}{3}+\emptyset)T_{1/\varepsilon+p}(\varepsilon)$$

By Robinson's lemma, there exists some unlimited $\nu > 0$ such that still $R_{1/\varepsilon-\nu-1}(\varepsilon) = (\nu-\frac{1}{3}+\emptyset)T_{1/\varepsilon-\nu}(\varepsilon)$. Then $T_{1/\varepsilon-\nu}(\varepsilon)/R_{1/\varepsilon-\nu-1}(\varepsilon) \simeq 0$.

3) The ratio R_{n-1}/T_n (i.e. the approximation factor associated to the expansion in question) is represented graphically in Fig. 1 above.

2. The approximation factor $A_\omega^f(x)$ near the convergence circle: case where $f^{(n)}$ is majorized.

Let $f(z) = \sum_{n=0}^{\infty} c_n z^n$ be a standard series such that $\lim_{n\to\infty} c_{n+1}/c_n = -1$. We know firstly, that outside the halo of the convergence circle $A_\omega^f(z) \simeq \frac{1}{1+z}$ and, secondly, that this approximation still must be true for $z \simeq 1$, if we wish to conclude that the Borel integral $F(\varepsilon) = \int_0^\infty e^{-t} f(\varepsilon t)$ is nearly optimally approximated by $\sum_{n=0}^{K} c_n n!\varepsilon^n$. In this section we show that $A_\omega^f(z)$ is approximately known near the convergence circle if $f^{(\omega)}$ is properly majorized.

We first consider a case where $f^{(\omega)}$ is known exactly. Let $f(z) = \frac{1}{(1+z)^r}$, $r \notin \mathbb{Z}^-$. The Taylor series of the function f is

$$f(z) = \sum_{n=0}^{\infty} (-1)^n \frac{(r)_n z^n}{n!}$$

The nth derivative of f is $f^{(n)}(z) = (-1)^n (r)_n (1+z)^{r-n}$. In Chapter I (proposition 1.6) we used this formula to prove that $A_\omega^f(z) \simeq \frac{1}{1+z}$ near the convergence circle. Indeed, we were able to approximate the integral

$$A_\omega^f(z) = \int_0^\omega \left(1-\frac{u}{\omega}\right)^{\omega-1} \frac{f^{(\omega)}\left(\frac{zu}{\omega}\right)}{f^{(\omega)}(0)} du$$

So Borel integrals $F(\varepsilon) = \int_0^\infty \frac{e^{-t}}{(1+\varepsilon t)^r} dt$ satisfy Theorem 3.2 (this can also be shown with the method of Chapter I, section 1).

Inspection of the proof of proposition 1.6 shows that it is not necessary to know $f^{(\omega)}$ exactly. Indeed, an approximation on the $1/\omega$-galaxy of 0 and a global majoration suffice.

The following proposition shows that an approximation of $f^{(\omega)}(\pounds/\omega)$ can be derived under fairly general conditions.

Notation: Let $n \in \mathbb{N}$. The function Ψ_n will be defined by

$$\Psi_n(u) = \frac{f^{(n)}(u/n)}{f^{(n)}(0)}.$$

Proposition 3.4: Let $f(z) = \sum_{n=0}^{\infty} c_n z^n$ be a standard series with nonzero convergence radius R. Suppose $(1/R)^{p(n)} = \limsup_{n\to\infty} |c_{n+p(n)}/c_n|$. Let ω be unlimited such that $c_\omega \neq 0$. Then for all limited u

$$\Psi_\omega(u) \simeq \sum_{n=0}^{\infty} \frac{a_n}{n!} u^n$$

where $(a_n)_{n\in\mathbb{N}}$ is the standard sequence defined by $a_n = {}^o(c_{\omega+n}/c_\omega)$, when st n.

<u>Proof</u>: (i) Let $u \in \mathbb{C}$ be limited. Because of the convergence one has $f^{(\omega)}(\frac{u}{\omega}) = \sum_{n=0}^{\infty} \frac{f^{(\omega+n)}(0)}{n!}(\frac{u}{\omega})^n$, so

$$\Psi_\omega(u) = \sum_{n=0}^{\infty} \frac{(\omega+1)\ \ldots\ (\omega+n)}{\omega^n} \frac{c_{\omega+n}}{c_\omega}.\frac{u^n}{n!}$$

Put $B_n = \frac{(\omega+1)\ \ldots\ (\omega+n)}{\omega^n} \frac{c_{\omega+n}}{c_\omega}.\frac{u^n}{n!}$. We approximate $\sum_{n=0}^{\infty} B_n$ with the aid of the lemma of dominated approximation. For all limited n

$$B_n \simeq {}^o(\frac{c_{\omega+n}}{c_\omega})\frac{u^n}{n!} = a_n \frac{u^n}{n!}$$

If $n \leq \omega$, then

$$|B_n| \leq \frac{2^n.(2/R)^n({}^o\ u\ +1)^n}{n!} = \frac{(4({}^o\ u\ +1)/R)^n}{n!}$$

If $n > \omega$, then from the inegality $\omega! \geq \omega^\omega e^{-2\omega}$

$$|B_n| \leq \frac{(\omega+1)\ \ldots\ (\omega+n)}{\omega^\omega}.\frac{(\omega+\omega+1)\ \ldots\ (\omega+n)}{(\omega+1)\ \ldots\ n}.\frac{(2/R)^n({}^o|u|+1)^n}{\omega!\omega^{n-\omega}}$$

$$\leq \frac{2^n(2({}^o|u|+1)/R)^n}{\omega^\omega\ e^{-2\omega}\ \omega^{n-\omega}} \quad (\frac{4e^2({}^o|u|+1)/R}{\omega})^n \leq (1/2)^n$$

Hence $|B_n|_{n\in\mathbb{N}}$ is majorized by the standard convergent sequence $\max((1/2)^n.\frac{(4({}^o|u|+1)/R)^n}{n!})$. So by the lemma of dominated approximation

$$\Psi_\omega(u) = \sum_{n=0}^{\infty} B_n \simeq \sum_{n=0}^{\infty} \frac{a_n}{n!}u^n$$

The last theorem of this section gives a growth condition on $f^{(\omega)}$ which is sufficient for Theorem 3.2 to hold.

<u>Theorem 3.5</u>: Let $f(z) = \sum_{n=0}^{\infty} c_n z^n$ be a standard series such that $\lim_{n\to\infty} c_{n+1}/c_n = -1$. Let $\omega \in \mathbb{N}$ be unlimited and suppose there exist st $L \in \mathbb{R}$ and st $B < 1$ such that $|f^{(\omega)}(z)| \leq |f^{(\omega)}(0)| L e^{\omega B|z|}$ for $|z| \lesssim 1$, $z \not\simeq -1$. Then for all $|z| \simeq 1$, $z \not\simeq -1$

$$A_\omega^f(z) \simeq \frac{1}{1+z}$$

Proof: One has

$$A_\omega^f(z) = {}_0\int^\omega (1-\frac{u}{\omega})^{\omega-1}\Psi_\omega(uz)du$$

Put $I(u) = (1-\frac{u}{\omega})^{\omega-1}\Psi_\omega(uz)$. By the preceding proposition one has for limited u

$$I(u) \simeq e^{-(1+z)u}$$

For all $u \leq \omega$ one has $I(u) \leq 2Le^{-(1-(|B|+1)/2)u}$. So, by the lemma of dominated approximation

$$\int_0^\omega I(u)du \simeq \int_0^\infty e^{-(1+z)u}du = \frac{1}{1+z}$$

Examples: Here we present formulas for the shadows (see page 162) of Ψ_ω and A_ω relative to some elementary functions.

f(z)	$({}^o\Psi_\omega)(z)$	$({}^oA_\omega)(z)$	
$\mathrm{Log}(1+z)$, $(1+z)^r$, $\Gamma(1+z)$ $r \notin \mathbb{N}$	e^{-z}	$\frac{1}{1+z}$	$z \notin]-\infty,-1]$
$(1+az^p)^r$ $a > 0$, $p \in \mathbb{N}$, $r \notin \mathbb{N}$ $\omega = p\nu$, $\nu \in \mathbb{N}$	$\sum_{n=0}^{\infty}(-1)^n \frac{a^n z^{pn}}{(pn)!}$	$\frac{1}{1+az^p}$	$\|z\| < 1$ or $\|z\| \geq 1$ and $\arg z \neq \frac{(2k-1)\pi}{p}$ $k \in \mathbb{Z}$
Arctg z ω even	cos z	$\frac{1}{1+z^2}$	$\|z\| < 1$ or $\|z\| \geq 1$ and $\mathrm{re}(z) \neq 0$
Arcsin z ω odd	cosh z	$\frac{1}{1-z^2}$	$\|z\| < 1$ or $\|z\| \geq 1$ and $\mathrm{im}(z) \neq 0$
e^z	1	1	

Notice that in these cases ${}^oA_\omega$ is the Borel transform of ${}^o\Psi_\omega$. Indeed, for standard z

$${}^o(A_\omega(z)) = {}^o({}_0\int^\omega (1-\frac{u}{\omega})^{\omega-1}\Psi_\omega(uz)du) = {}_0\int^\infty e^{-u}({}^o\Psi_\omega)(uz)du$$

3. The approximation factors A_ω^φ, A_ω^f and A_ω^F when $c_{n+p}/c_n = c+o(1/\sqrt{n})$

If the sequence of coefficients $(c_n)_{n\in\mathbb{N}}$ satisfies the property $c_{n+p}/c_n = c+o(1/\sqrt{n})$ for $n \to \infty$, instead of just $c_{n+p}/c_n \to c$ for $n \to \infty$, the situation is much more regular. Indeed, it is now possible to extend the validity of the approximation of the approximation factors A_ω^φ, A_ω^f and A_ω^F to the "critical regions". In particular this means that the property of the "summation to the nearly minimal term" holds for the divergent expansion $F(z) \sim \sum_{n=0} c_n n! z^n$.

This section has the following structure. First we investigate the stability of the condition $c_{n+p}/c_n = c+o(1/\sqrt{n})$ under multiplication of power series. Secondly we mention a result by Polya, which relates the growth of $\varphi(z) = \sum_{n=0}^{\infty} c_n \frac{z^n}{n!}$ and the singularities of its Borel transform $f(z) = \sum_{n=0}^{\infty} c_n z^n$. This result will be needed later on. Thirdly we present a global approximation of A_ω^φ, which is established in a direct, but elaborate way. Taking Borel transforms, we obtain an approximation of A_ω^f near the convergence circle. Again taking Borel transforms we obtain an approximation of A_ω^F for indices near that of the smallest term.

Stability of the condition.

Theorem 3.6 is the counterpart of Theorem 2.11 for the stability of the condition $c_{n+p}/c_n = c+o(1/\sqrt{n})$ under multiplication of power series. For simplicity of notation, we consider the case p=1.

Theorem 3.5: Let $f(z) = \sum_{n=0}^{\infty} a_n z^n$ and $g(z) = \sum_{n=0}^{\infty} b_n z^n$ be two analytic functions such that $a_{n+1}/a_n = a+o(1/\sqrt{n})$, with $a \neq 0$ and g has convergence radius $R > |1/a|$. Let $\sum_{n=0}^{\infty} c_n z^n$ be the power series of f.g. Then

(i) $\quad c_n/a_n = g(1/a)+o(1/\sqrt{n})$

(ii) $\quad$ if $g(1/a) \neq 0$, then $c_{n+1}/c_n = a+o(1/\sqrt{n})$

Proof: (i) By Transfer, f and g may be supposed standard. Let $\omega \in \mathbb{N}$ be unlimited. One has

$$c_\omega = a_\omega(b_0+b_1\frac{a_{\omega-1}}{a_\omega} + \dots + a_1\frac{b_{\omega-1}}{a_\omega} + a_0\frac{b_\omega}{a_\omega})$$

Put $d_n = b_n \frac{a_{\omega-n}}{a_\omega}$ and $h(z) = \sum_{n=0}^{\infty} |b_n| z^n$. From the first part of the proof of Theorem 2.2, we may conclude that there exists unlimited ν (which may be assumed $\geq \omega/2$) such that

$$\sum_{n=\nu+1}^{\omega} d_n \in 1/\omega\text{-}M$$

For $n \leqq \omega-\nu$ one has the estimation

$$\frac{a_{\omega-n}}{a_\omega} = \frac{a_{\omega-1}}{a_\omega}\cdot\frac{a_{\omega-2}}{a_{\omega-1}}\cdot \ldots \cdot \frac{a_{\omega-n}}{a_{\omega-n+1}}$$

$$= (1/a + \frac{\emptyset}{\sqrt{\omega-1}})(1/a + \frac{\emptyset}{\sqrt{\omega-2}}) \cdot \ldots \cdot (1/a + \frac{\emptyset}{\sqrt{\omega-n}})$$

$$= (1/a)^n e^{\sum_{k=1}^{n} \log(1+\frac{\emptyset}{\sqrt{\omega-k}})} = (1/a)^n e^{\sum_{k=1}^{n} \frac{\emptyset}{\sqrt{\omega-k}}}$$

$$= (1/a)^n e^{\emptyset \int_1^n \frac{1}{\sqrt{\omega-x}}dx} = (1/a)^n e^{\emptyset(\sqrt{\omega-n}-\sqrt{\omega-1})} = (1/a)^n e^{(\alpha_n/\sqrt{\omega})n}$$

$$(\alpha_n \simeq 0)$$

Hence, with $\alpha = \max_{n \leqq \nu} |\alpha_n|$

$$|\sum_{n=0}^{\nu} d_n - g(1/a)| = |\sum_{n=0}^{\nu} b_n (1/a)^n - g(1/a) + \sum_{n=0}^{\nu} b_n/a^n (e^{(\alpha_n/\sqrt{\omega})n} - 1)|$$

$$\leqq \mu_2 + \sum_{n=0}^{\nu} |b_n|/|a|^n (e^{(\alpha/\sqrt{\omega})n} - e^{-(\alpha/\sqrt{\omega})n}) \qquad (\mu_2 \in 1/\omega^{-M})$$

$$= h(\frac{e^{\alpha/\sqrt{\omega}}}{|a|}) - h(\frac{e^{-\alpha/\sqrt{\omega}}}{|a|}) + \mu_3 \qquad (\mu_3 \in 1/\omega^{-M})$$

$$= \frac{2\alpha}{|a|\sqrt{\omega}}(h'(1/|a|)+\emptyset) + \mu_3 = \frac{\emptyset}{\sqrt{\omega}}$$

Combining, $c_\omega = a_\omega(g(1/a)+\emptyset/\sqrt{\omega})$, as required.

(ii) One has

$$\frac{c_{\omega+1}}{c_\omega} = \frac{a_{\omega+1}}{a_\omega}(\frac{g(1/a)+\emptyset/\sqrt{\omega+1}}{g(1/a)+\emptyset/\sqrt{\omega}}) = (a+\emptyset/\sqrt{\omega})(1+\emptyset/\sqrt{\omega}) = a+\emptyset/\sqrt{\omega}$$

A wide family of functions satisfying the property $c_{n+1}/c_n = 1+o(1/\sqrt{n})$ for $n \to \infty$ is formed by the <u>hypergeometric functions</u>

$$F(a,b;c;z) = \sum_{n=0}^{\infty} \frac{(a)_n (b)_n}{(c)_n} \frac{z^n}{n!}$$

One even has $c_{n+1}/c_n = 1+O(1/n)$ for $n \to \infty$.

<u>The growth of $\varphi(z)$.</u>

Let again $f(z) = \sum_{n=0}^{\infty} c_n z^n$ where we assume for reasons of simplicity that

$c_{n+1}/c_n = -1+o(1/\sqrt{n})$. In Theorem 3.12 below we show that $A_\omega^f(z) \simeq \frac{1}{1+z}$ in the halo of the convergence circle by taking the Borel transform of $\varphi(z) = \sum_{n=0}^{\infty} \frac{c_n z^n}{n!}$:

$$f(z) = \int_0^\infty e^{-t}\varphi(zt)dt$$

In order to do so we need that $\varphi(z)$ does not grow to fast in the directions under consideration. In particular we will ask that $|\varphi(z)| \leq Ke^{B|z|}$ where K, B are standard and $B < 1$. Now it is very difficult to read the growth of $\varphi(z)$ directly from its Taylorseries, when the c_n are not ultimately of the same sign. However, it is possible to relate the growth of φ to the position of the singularities of f, by means of a theorem of Polya. We use the following definitions (see Henrici [47]).

Definition 3.7: Let φ be an entire function of exponential order and Θ a direction in the complex plane. Then the function $\Gamma(\Theta)$, defined by

$$\Gamma(\Theta) = \inf\{a \mid (\exists b) \mid \varphi(re^{i\Theta})| \leq e^{ar} \text{ for all } r \geq b\}$$

is called the indicator of φ.
For instance if $\varphi(z) = e^z$, then $\Gamma(0) = \cos\Theta$, and if $\varphi(z) = \cos z$, then $\Gamma(\Theta) = |\sin\Theta|$.

Definition 3.8: Let C be a bounded closed convex subset of $\mathbb{C}$ and Θ a direction in the complex plane. Then the function $K(\Theta)$ defined by

$$K(\Theta) = \sup_{z \in C} \operatorname{re}(e^{-i\Theta}z)$$

is called the support function of C. If C is the convex closure of the singularities of a function of a function f we write $K = K_f$.

For instance, if $C = \{1\}$, then $K(\Theta) = \cos\Theta$, and if $C = \{i,-i\}$ then $K(\Theta) = |\sin\Theta|$. Also if C is contained in the half plane $|z| < 1$, then $K(0) < 1$; if $C = [-1,0]$, then $K(\Theta) = -\cos\Theta$ for $\frac{\pi}{2} \leq \Theta \leq \frac{3\pi}{2}$ and $K(\Theta) = 0$ for $-\frac{\pi}{2} \leq \Theta \leq \frac{\pi}{2}$.

Notice that for analytic functions f with real coefficients K_f is symmetric with respect to the real axis, i.e. $K_f(\Theta) = K_f(-\Theta)$.

Theorem 3.9: (Polya): Let φ be an entire function of exponential order. Let $\Gamma(\Theta)$ be its indicator and $g(z) = \int_0^\infty e^{-zt}\varphi(t)dt$ be its Laplace transform. Let $K_g(\Theta)$ be the support function of the singularities of g. Then

$$\Gamma(\Theta) = K_g(-\Theta)$$

The Borel transform f and the Laplace transform are related by

$$g(z) = \frac{1}{z}\cdot f(\frac{1}{z})$$

Hence the corresponding formula for Borel transforms reads

$$\Gamma(\Theta) = K_{\frac{1}{z}.f(\frac{1}{z})}(-\Theta)$$

Notice that the formula reduces to $\Gamma(\Theta) = K_{\frac{1}{z}.f(\frac{1}{z})}(\Theta)$ if f has real coefficients. Let us check this formula on some easy examples. If $\varphi(z) = e^z$, then $f(z) = \frac{1}{1-z}$. Indeed, $K_{\frac{1}{z}.f(\frac{1}{z})}(\Theta) = \Gamma(\Theta) = \cos\Theta$. If $\varphi(z) = \cos z$, then $f(z) = \frac{1}{1+z^2}$. Here $K_{\frac{1}{z}.f(\frac{1}{z})}(\Theta) = \Gamma(\Theta) = |\sin\Theta|$.

The following cases are important to us. (Case (1)). Suppose that all singularities of f are contained in the strip $]-\infty,-1]$. Then the singularities of $\frac{1}{z}.f(\frac{1}{z})$ are contained in $[-1,0]$. So $\Gamma(\Theta) = -\cos\Theta$ for $\frac{\pi}{2} \leq \Theta \leq \frac{3}{2}\pi$ and $K(\Theta) = 0$ for $-\frac{\pi}{2} \leq \Theta \leq \frac{\pi}{2}$. Hence $|\varphi(z)| \leq Ke^{B|z|}$, where st $B < 1$, in all directions $\not\simeq \pi$.
(Case (2)). Suppose 1 is not a singularity of f. Then there is no singularity in some closed disk of the form $B(\frac{1}{2},\frac{1}{2}+\alpha)$ with $d > 1$. So all singularities of $\frac{1}{z}.f(\frac{1}{z})$ lie in the half plane re $z \leq \frac{1}{1+d}$. Then $|\varphi(z)| \leq Ke^{B|z|}$ with st K and st $B < 1$ at least in the direction of the positive real axis.

In the direction of the negative real axis the growth of $\varphi(z)$ can be determined directly. See the excercise.

<u>Excercise</u>: Let $\varphi(x) = \sum_{n=0}^{\infty} \frac{c_n x^n}{n!}$, where $x > 0$ and $c_{n+1}/c_n = 1+o(1/\sqrt{n})$ for $n \to \infty$.

1) Prove that for all unlimited $\omega > 0$

$$\varphi(\omega) = (1+\emptyset)\sqrt{2\pi\omega}\, c_{[\omega]} \frac{\omega^\omega}{\omega!}$$

Hint: use the method of proposition 3.10 below. It is not needed to apply summation by parts.

2) Use 1) to give an alternative proof of Stirling's formula.

<u>The remainder $\rho_\omega(z)$ in the critical region.</u>

Let $\varphi(z) = \sum_{n=0}^{\infty} \frac{c_n}{n!}z^n$, where we assume as usual that $c_{n+1}/c_n = -1+o(1/\sqrt{n})$ for $n \to \infty$ and let $\omega \in \mathbb{N}$ be unlimited. By Theorem 2.14 the remainder $\rho_\omega(z)$ is approximately known for $|z|/\omega \not\simeq 1$. We will now derive an approximation for $\rho_\omega(z)$ in the "critical region" $|z|/\omega \simeq 1$, with the exception of the halo $z/\omega \simeq -1$. The proof is straightforward, but uses some careful estimations. It is based on Abel's partial summation formula.

<u>Abel's partial summation formula:</u> Let $(a_n)_{n\in\mathbb{N}}$ and $(b_n)_{n\in\mathbb{N}}$ be two sequences. Put

$$A_n = \sum_{k=0}^{n} a_k$$

if $n \geq 0$. Put $A_{-1} = 0$. Then, if $0 \leq p \leq q$, we have

$$\sum_{n=p}^{q} a_n b_n = \sum_{n=p}^{q-1} A_n(b_n-b_{n+1})+A_q b_q - A_{p-1}b_p.$$

<u>Notations</u>. Let $\phi(z)$ be an entire function and $\omega \in \mathbb{N}$ be unlimited. Put

$$\Theta = \arg z$$

$$A_k = \frac{c_{\omega+k}}{c_\omega}\,\frac{|z/\omega|^k}{(1+1/\omega).\ \dots\ .(1+k/\omega)}$$

$$B_k = A_k - A_{k+1}$$

$$C_k = B_k - B_{k+1}$$

$$D_k = \frac{c_{\omega-1-k}}{c_{\omega-1}}(1-1/\omega).\ \dots\ .(1-k/\omega)\,|\omega/z|^k$$

With these notations the following identities hold:

(1) $$\rho_{\omega-1}(z) = \tau_\omega(z)\sum_{k=0}^{\infty} e^{i\Theta k}A_k$$

(2) $$\rho_{\omega-1}(z) = \varphi(z)-\tau_{\omega-1}(z)\sum_{k=0}^{\infty} e^{-i\Theta k}D_k$$

(3) $$\sum_{k=0}^{\nu} e^{i\Theta k}A_k = \frac{1}{1-e^{i\Theta}}\sum_{k=0}^{\nu-1} B_k - \frac{e^{i\Theta}}{(1-e^{i\Theta})^2}\sum_{k=0}^{\nu-2}(1-e^{i\Theta(k+1)})C_k + \frac{1-e^{i\Theta\nu}}{1-e^{i\Theta}}A_\nu - \frac{e^{i\Theta}-e^{i\Theta\nu}}{(1-e^{i\Theta})^2}B_{\nu-1}$$

(by Abel's partial summation formula, applied two times)

(4) $$A_k = \frac{c_{\omega+k}}{c_\omega} e^{-\sum_{m=1}^{k} \log(1+m/\omega)} |z/\omega|^k$$

(5) $$B_k = A_k\left(1 - \frac{c_{\omega+k+1}}{c_{\omega+k}} \frac{|z/\omega|}{1+(k+1)/\omega}\right)$$

(6) $$C_k = A_k\left(1 - \frac{2c_{\omega+k+1}}{c_{\omega+k}} \frac{|z/\omega|}{1+(k+1)/\omega} + \frac{c_{\omega+k+2}}{c_{\omega+k}} \frac{|z/\omega|^2}{(1+(k+1)/\omega)(1+(k+2)/\omega)}\right)$$

If $S \subset R$ is a set, then we write

$$x < S \Leftrightarrow x < s \text{ for all } s \in S$$

$$x \leqq S \Leftrightarrow \exists s \in S \text{ such that } x \leqq s$$

Proposition 3.10: Let $\varphi(z) = \sum_{n=0}^{\infty} \frac{c_n z^n}{n!}$, where $c_{n+1}/c_n = 1+o(1/\sqrt{n})$ for $n \to \infty$. Let $\omega \in \mathbb{N}$ be unlimited, $|z/\omega| \simeq 1$ and $\Theta \not\simeq 0$. Then

(i) if $|z| \leqq \omega+\sqrt{\omega}.\text{gal}$, then $\rho_{\omega-1}(z) = (1+\emptyset)\frac{\tau_\omega(z)}{1-e^{i\Theta}}$

(ii) if $|z| > \omega+\sqrt{\omega}.\text{gal}$, then $\rho_{\omega-1}(z) = (1+\emptyset)\frac{\tau_\omega(z)}{1-e^{i\Theta}} + \varphi(z)$

Proof:(i) The idea of the proof is as follows: We use the formula $\rho_{\omega-1}(z) = \sum_{n=\omega}^{\infty} \frac{c_n z^n}{n!}$. In estimating this sum, essentially only the terms of index $(1+\emptyset)|z|$ matter. But these terms wipe each other almost out. We use Abel's partial summation formula to bypass this difficulty.

So we use identity (1). Let $|z| = \omega(1+\alpha)$, where $\alpha \simeq 0$. If $b \gtrsim 0$, then $A_{b\omega}$, $B_{b\omega}$ and $\sum_{k=b\omega}^{\infty} e^{i\Theta k}$ all belong to the microhalo $1/\omega$-M; indeed, it is easily verified that they are of order $e^{-d\omega}$ with $d \gtrsim 0$. By the Fehrele principle, there exists $\eta \simeq 0$ such that A_k, B_{k-1} and $\sum_{k=\eta\omega}^{\infty} e^{ik\Theta} A_k$ belong to $1/\omega$-M for all $k \geqq \eta\omega$. Then by the identity (3)

$$\sum_{k=0}^{\infty} e^{i\Theta k} A_k = \sum_{k=0}^{\eta\omega} e^{i\Theta k} A_k + \mu_1 \qquad (\mu_1 \in 1/\omega\text{-M})$$

$$= \frac{1}{1-e^{i\Theta}} \sum_{k=0}^{\eta\omega-1} B_k - \frac{e^{i\Theta}}{(1-e^{i\Theta})^2} \sum_{k=0}^{\eta\omega-2} (1-e^{i\Theta(k+1)})C_k + \mu_2 \quad (\mu_2 \in 1/\omega\text{-M})$$

The sum $\sum_{k=0}^{\eta\omega-1} B_k$ will be approximated by an integral using the following approximations, valid for all k such that $k/\omega \simeq 0$. They are obtained from the identities (4) and (5).

(7) $$A_k = e^{\emptyset . k/\sqrt{\omega}-(1+\emptyset)k^2/2\omega+(1+\emptyset)k\alpha}$$

(8) $$B_k = A_k(\emptyset/\sqrt{\omega}-\alpha+k/\omega)(1+\emptyset)$$

The sum $|\sum_{k=0}^{\eta\omega-2}(1-e^{i\Theta(k+1)})C_k|$ will be majorized by an integral. Indeed, for all $k \leq \eta\omega-2$, from the identity (6) we obtain the approximation

$$C_k = A_k(\emptyset/\sqrt{\omega}-\emptyset.\alpha+\emptyset.k^2/\omega^2)$$

So there exists $\gamma \simeq 0$ such that for all $k \leq \eta\omega-2$

$$|C_k| \leq \gamma.A_k(1/\sqrt{\omega}+|\alpha|+k^2/\omega^2)$$

From now on we distinguish two cases: a. $\alpha\sqrt{\omega}$ unlimited, b. $\alpha\sqrt{\omega}$ limited. It will be seen that the order of magnitude of α influences the integration step.

a. $\alpha\sqrt{\omega}$ unlimited: Let us write $\delta = -\alpha$. Then $z = \omega(1-\delta)$, where $\delta > 0$ and $\delta\sqrt{\omega}$ is unlimited. Here $\{k \in \mathbb{N} \mid A_k \not\simeq 0\} = 1/\delta\text{-gal} \cap \mathbb{N}$. This suggests to let δ be the integration step. Put

$$G_n = \sum_{k=[n/\delta]}^{[(n+1)/\delta]-1} B_k$$

For all limited n one has the approximation

$$G_n = \sum_{k=[n/\delta]}^{[(n+1)/\delta]-1} e^{-\delta k(1+\emptyset)}\delta(1+\emptyset) = (1+\emptyset)\sum_{k=[n/\delta]}^{[(n+1)/\delta]-1} e^{-\delta k}\delta \simeq \int_n^{n+1} e^{-t}dt$$

Let $\nu \in \mathbb{N}$ be minimal such that $[(\nu+1)/\delta] \geq \eta\omega$. From (7) and (8) one obtains for all $n \leq \nu$

$$G_n \leq \sum_{k=[n/\delta]}^{[(n+1)/\delta]-1} e^{-\delta k/2}(2\delta+\frac{2}{\omega\delta^2}.\delta k.\delta) \leq \int_n^{n+1} e^{-t/2}(2+t)dt$$

Hence by the lemma of dominated approximation

$$\sum_{n=0}^{\nu} G_n \simeq \int_0^{\infty} e^{-t}dt = 1$$

In the same manner one proves that $|\sum_{k=0}^{[(\nu+1)/\delta]-2}(1-e^{i\Theta(k+1)})C_k| \leq 2\gamma\int_0^{\infty} e^{-t/2}(2+t^2)dt \simeq 0$. Hence

$$\sum_{k=0}^{\eta\omega} e^{i\Theta k}A_k = \sum_{k=0}^{[(\nu+1)/\delta]} e^{i\Theta k}A_k+\mu_3 \simeq \frac{1}{1-e^{i\Theta}} \qquad (\mu_3 \in 1/\omega\text{-M})$$

This implies that $\rho_{\omega-1}(z) = (1+\emptyset)\dfrac{\tau_\omega(z)}{1-e^{i\Theta}}$.

b. $\underline{a\sqrt{\omega}\ \text{limited}}$: Let us write $z = \omega + a\sqrt{\omega}$, where a is limited. Here $\{k \in \mathbb{N} \mid A_k \not\simeq 0\} = 1/\sqrt{\omega}\text{-gal} \cap \mathbb{N}$. This suggests to let $1/\sqrt{\omega}$ be the integration step. Put

$$H_n = \sum_{k=[n\sqrt{\omega}]}^{[(n+1)\sqrt{\omega}]-1} B_k$$

For all limited n one has the approximation

$$H_n = \sum_{k=[n\sqrt{\omega}]}^{[(n+1)\sqrt{\omega}]-1} e^{\emptyset k/\sqrt{\omega}-(1+\emptyset)(k/\sqrt{\omega})^2/2+(1+\emptyset)ka/\sqrt{\omega}}(-a+\emptyset+k/\sqrt{\omega})(1+\emptyset)/\sqrt{\omega}$$

$$= (1+\emptyset)\sum_{k=[n\sqrt{\omega}]}^{[(n+1)\sqrt{\omega}]-1} e^{-(k/\sqrt{\omega})^2/2+ka/\sqrt{\omega}}(-a+k/\sqrt{\omega})/\sqrt{\omega}.$$

$$= \int_n^{n+1} e^{-t^2/2+at}(-a+t)dt$$

Let $\nu \in \mathbb{N}$ be minimal such that $[(\nu+1)\sqrt{\omega}]-1 \geqq \eta\omega$. From (7) and (8) one obtains for all $n \leqq \nu$

$$|H_n| \leqq \int_n^{n+1} e^{-t^2/4+(|a|+1)t}(|a|+1+2t)dt$$

Hence by the lemma of dominated approximation

$$\sum_{n=0}^{\nu} H_n \simeq \int_0^\infty e^{-t^2/2+at}(-a+t)dt = 1$$

In the same manner one proves that $\left|\sum_{k=0}^{[(\nu+1)\sqrt{\omega}]-2} (1-e^{i\Theta(k+1)})C_k\right| \leqq 2\gamma \int_0^\infty e^{-t^2/4+(|a|+1)t}(|a|+1+t^2)dt \simeq 0$. Hence

$$\sum_{k=0}^{\eta\omega} e^{i\Theta k}A_k = \sum_{k=0}^{[(\nu+1)/\sqrt{\omega}]} e^{i\Theta k}A_k + \mu_4 \simeq \frac{1}{1-e^{i\Theta}} \qquad (\mu_4 \in 1/\omega\text{-M})$$

This implies that $\rho_{\omega-1}(z) = (1+\emptyset)\dfrac{\tau_\omega(z)}{1-e^{i\Theta}}$

(ii) $\underline{|z| > \omega+\sqrt{\omega}\text{-gal}}$: The proof is analogous to (i)$\underline{a}$, starting from formula (2).

As a corollary we obtain the following global (that is to say, with the exception of the external sector $\text{Arg } z \simeq \pm\pi$) approximation of $\rho_{\omega-1}(z)$.

<u>Theorem 3.11</u>: Let $\varphi(z) = \sum_{n=0}^{\infty} \frac{c_n z^n}{n!}$, where $c_{n+1}/c_n = -1+o(1/\sqrt{n})$ for $n \to \infty$. Let $\omega \in \mathbb{N}$ be unlimited and $\text{Arg } z \not\simeq \pm\pi$. Then there exists unlimited $\sigma > 0$ such that

(i) if $|z| \lessapprox \omega+\sigma\sqrt{\omega}$, then $\rho_{\omega-1}(z) = (1+\emptyset)\frac{\tau_\omega(z)}{1+\omega/z}$

(ii) if $|z| > \omega+\sigma\sqrt{\omega}$, then $\rho_{\omega-1}(z) = (1+\emptyset)\frac{\tau_\omega(z)}{1+\omega/z} + \varphi(z)$

Proof: The assertions follow from theorem 2.14, proposition 3.10 and an application of the Fehrele principle.

Application to the problem of the "summation to the smallest term".

Let $f(z) = \sum_{n=0}^{\infty} c_n z^n$, where $c_{n+1}/c_n = -1+o(1/\sqrt{n})$ for $n \to \infty$. We use Theorem 3.11 to extend the approximation $\phi_\omega{}^f \simeq \frac{1}{1+z}$ to the halo of the convergence circle.

Theorem 3.12: Let f(z) be the analytic continuation of a series $\sum_{n=0}^{\infty} c_n z^n$, where $c_{n+1}/c_n = -1+o(1/\sqrt{n})$ for $n \to \infty$. Let $\omega \in \mathbb{N}$ be unlimited and $\Theta = \operatorname{Arg} z \not\simeq \pm\pi$. Suppose there exist standard constants K and B, with B < 1 such that $|\varphi(z)| = |\sum_{n=0}^{\infty} \frac{c_n}{n!} z^n| \lessapprox Ke^{B|z|}$ in the direction Θ. Then

$$A_\omega{}^f(z) \simeq \frac{1}{1+z}$$

for all $|z| \simeq 1$ such that $\operatorname{Arg} z = \Theta$.

Proof: Let $\sigma > 0$ be unlimited, as in theorem 3.11. Then

$$\begin{aligned}
r_{\omega-1}(z) &= \int_0^\infty e^{-t}\rho_{\omega-1}(zt)dt \\
&= \int_0^\infty e^{-t}\frac{\tau_\omega(zt)}{1+zt/\omega}(1+\alpha(t)dt + \int_{(\omega+\sigma\sqrt{\omega})/|z|}^\infty e^{-t}\varphi(zt)dt \qquad (\alpha(t) \simeq 0) \\
&= \frac{c_\omega z^\omega}{\omega!}\int_0^\infty e^{-t}\frac{t^\omega}{1+zt/\omega}(1+\alpha(t)dt + \Theta_1 K \int_{(\omega+\sigma\sqrt{\omega})/|z|}^\infty e^{-t(1-B|z|)}dt \\
& \qquad\qquad (|\Theta_1| \lessapprox 1) \\
&= (1+\emptyset)\frac{c_\omega z^\omega}{1+z} + \Theta_2 e^{-d\omega} \qquad (|\Theta_2| = 1,\ d \gtrapprox 0) \\
&= (1+\emptyset)\frac{t_\omega(z)}{1+z}
\end{aligned}$$

The last identity is a consequence of the estimation $|c_\omega z^\omega| = e^{\emptyset\omega}$.

As a consequence, we derive the property of the summation to the nearly minimal term for approximations of Borel integrals $\int_0^\infty e^{-t}f(\varepsilon t)dt$ by the series $\sum_{n=0} c_n n!\varepsilon^n$, for the case $c_{n+1}/c_n = -1+o(1/\sqrt{n})$ and $\varepsilon \simeq 0$, $\operatorname{Arg} \varepsilon \not\simeq \pm\pi$.

<u>Theorem 3.13</u>: Let $f(z)$ be the analytic continuation of a series $\sum_{n=0}^{\infty} c_n z^n$ where $c_{n+1}/c_n = -1+o(1/\sqrt{n})$. Let $\varepsilon \simeq 0$ be such that st Arg ε, Arg $\varepsilon \neq \pm\pi$ and f is of exponential order in this direction. Let $F(\varepsilon)$, $T_n(\varepsilon)$ and $R_{n-1}(\varepsilon)$ be as usual. Then

(i) For all $n \in \mathbb{N}$ one has $R_{n-1}(\varepsilon) = (1+\emptyset)\frac{T_n(\varepsilon)}{1+n\varepsilon}$

(ii) A remainder $R_k(\varepsilon)$ is nearly minimal if and only if the term $T_k(\varepsilon)$ is nearly minimal.

<u>Proof</u>: Put Θ = Arg ε. Let $\omega \in \mathbb{N}$ be unlimited. We show that $A_\omega^f(z) \simeq \frac{1}{1+z}$ for all limited z such that Arg $z = \Theta$, then all conditions of Theorem 3.2 are fulfilled. By Theorem 2.15 the approximation $A_\omega^f(z) \simeq \frac{1}{1+z}$ holds for all limited z in the direction Θ such that $|z| \not\simeq 1$. Define

$$h(w) = f(we^{i\Theta})$$

It is supposed that f can be continuated analytically along the line Arg $z = \Theta$, so the function h has no singularities for w real and positive. Then Theorem 3.9 implies (see case (2), p. 85) that $|\varphi(we^{i\Theta})| \leqq Ke^{Bw}$ for w real and positive, where st K, B and $B < 1$. Hence $|\varphi(z)| \leqq Ke^{B|z|}$ for Arg $z = \Theta$. Then Theorem 3.12 implies that $A_\omega(z) \simeq \frac{1}{1+z}$ for all $z \simeq e^{i\Theta}$. Hence $A_\omega^f(z) \simeq \frac{1}{1+z}$ for all limited z such that Arg $z = \Theta$. Now the assertions follow from Theorem 3.2.

We return to the property of summation to the (nearly) minimal term in a more theoretical context in the last section of chapter VI.

4. <u>Application: Uniform behaviour of remainders of high order.</u>

The results of this and the preceding chapter imply that under fairly general conditions the remainders of high order $R_\omega(x)$ of Taylor expansions or asymptotic expansions of analytic functions present a rather uniform behaviour: they increase exponentially with x in a way which depends less on the individual function than on the ratio x/ω.

Indeed, let $G(x) = \sum_{n=0}^{\infty} a_n x^n$ be a standard analytic function. Let $\omega \in \mathbb{N}$ be unlimited and $x \geqq 0$. We consider $R_{\omega-1}^G(x)$ as a function of x. We have the formula

$$R_{\omega-1}^G(x) = T_\omega^G(x).A_\omega^G(x)$$

Now as a function of x the term $T_\omega^G(x)$ increases very rapidly, for $T_\omega^G(x)$ is a constant multiplied by x^ω. On the contrary the approximation factor $A_\omega^G(x)$ often

x	$R_n(x)$ for e^x	$R_n(x)$ for $\frac{1}{1+x}$	
0.1	1.061942385E-261	-9.090909090E-102	
0.2	2.694988467E-231	-2.112751000E-071	
0.3	1.645132494E-213	-1.189332740E-053	
0.4	6.846056738E-201	-4.581251555E-041	
0.5	4.205130100E-191	-2.629536350E-031	
0.6	4.183245277E-183	-2.449944838E-023	
0.7	2.418618433E-176	-1.331843268E-016	
0.8	1.742541193E-170	-9.053493228E-011	
0.9	2.558686673E-165	-1.258171526E-005	
1.0	1.071404229E-160	-5.000000000E-001	n=100
1.1	1.625715989E-156	-7.218415987E+003	
1.2	1.066890541E-152	-4.517344064E+007	
1.3	3.463566683E-149	-1.401363323E+011	
1.4	6.174587507E-146	-2.391775218E+014	
1.5	6.566369347E-143	-2.439367065E+017	
1.6	4.453060633E-140	-1.589076848E+020	
1.7	2.033827236E-137	-6.981959013E+022	
1.8	6.545292773E-135	-2.164536851E+025	
1.9	1.541534483E-132	-4.917175378E+027	
2.0	2.743487717E-130	-8.451004001E+029	

0.1	2.526249341E-019	-9.090909091E-012	
0.2	5.217517598E-016	-1.706666667E-008	
0.3	4.551469803E-014	-1.362669231E-006	
0.4	1.086896308E-012	-2.995931429E-005	
0.5	1.276248878E-011	-3.255208333E-004	
0.6	9.565183202E-011	-2.267481600E-003	
0.7	5.258956960E-010	-1.163133378E-002	
0.8	2.304785065E-009	-4.772185884E-002	
0.9	8.494849217E-009	-1.651634718E-001	
1.0	2.731266076E-008	-5.000000000E-001	n=10
1.1	7.863000066E-008	-1.358627003E+000	
1.2	2.066314048E-007	-3.377310776E+000	
1.3	5.029709266E-007	-7.792001713E+000	
1.4	1.147011193E-006	-1.697210021E+001	
1.5	2.472786447E-006	-3.459902344E+001	
1.6	5.076409958E-006	-6.766225402E+001	
1.7	9.982978326E-006	-1.269329493E+002	
1.8	1.889887923E-005	-2.295300360E+002	
1.9	3.458409241E-005	-4.016905479E+002	
2.0	6.138893594E-005	-6.826666667E+002	

Figure 2: The remainders $R_n(x)$ associated to the Taylor expansions $e^x = \sum_{k=0}^{\infty} \frac{x^k}{k!}$ and $\frac{1}{1+x} = \sum_{k=0}^{\infty} (-1)^k x^k$ for x=0.1 to x=2 and for n=100 (above) and n=10 (below). Though the convergence properties of both series are different (the first is everywhere convergent, the second possesses a convergence circle, with radius 1), the behaviour of the remainders as a function of x is sensibly the same: for x=100 their proportion is about 10^{-159}, for n=10 their proportion is about 10^{-7}.

behaves very gently as a function of x: in most of the cases we treated $A_\omega^G(x)$ was S-continuous on a large domain. For instance, if $\lim_{n\to\infty} \frac{c_{n+1}}{c_n} = -1$, and if we define as usual $\varphi(x) = \sum_{n=0}^{\infty} \frac{c_n x^n}{n!}$ and $f(x) = \sum_{n=0}^{\infty} c_n x^n$, then

$$A_\omega^\varphi(x) = \frac{1+\emptyset}{1+x/\omega} \qquad x/\omega \lneqq 1$$

$$A_\omega^f(x) = \frac{1+\emptyset}{1+x} \qquad x \text{ limited, } |x| \not\simeq 1\text{, and } x \text{ not nearly equal to any singularity}$$

The results of this chapter show that the domains where these approximations are valid can often be extended and then the approximation factor is S-continuous on a larger domain. Also the approximation factor may be S-continuous without the condition $\lim_{n\to\infty} \frac{c_{n+1}}{c_n} = -1$ being fulfilled. For example, the condition may be weakened to $0 < m \leqq \left|\frac{c_{n+p(n)}}{c_n}\right| \leqq M < \infty$. Then proposition 2.24 shows that the approximation factor $A_\omega^f(x)$ is infinitely close to a standard continuous function, and thus S-continuous, within the convergence disk, with the possible exclusion of the halo of its border.

We conclude that if $A_\omega^G(x)$ is S-continuous on some interval $[0,\bar{x}]$ the remainder $R_{\omega-1}^G(x)$ essentially depends on the factor x^ω. So if $H(x) = \sum_{n=0}^{\infty} b_n x^n$ is another analytic function such that $A_\omega^H(x)$ is S-continuous on $[0,\bar{x}]$, the evolution of $R_{\omega-1}^H(x)$ and $R_{\omega-1}^G(x)$ as a function of x is very much the same. More precisely, if we suppose in addition that $\bar{x}$ is limited and that both $A_\omega^G(x)$ and $A_\omega^H(x)$ are appreciable on $[0,\bar{x}]$, then there exists a constant K ($= a_\omega/b_\omega$) and a standard continuous non zero function χ ($= {}^oA_\omega^G/{}^oA_\omega^H$) ("o" means "shadow of", see page 162) such that for all $x \in [0,\bar{x}]$

$$R_{\omega-1}^G(x) = (1+\emptyset)K\chi(x)R_{\omega-1}^H(x)$$

In particular the behaviour of a remainder is in this situation not affected by a possible transition from the domain of convergence to the domain of divergence of the series. One might even say that if the remainder is known approximately in some point of the interval $[0,\bar{x}]$ its order of magnitude is known everywhere on the interval (one could take $R_{\omega-1}^{\exp}$ as a standard). See Fig. 2.

The uniform behaviour of remainders of high order is very outspoken, if we consider them on not too long distances. Indeed, for all analytic functions G such that $A_\omega^G(x)$ is limited on $[0,\bar{x}]$ we have, as long as $x \lneqq \bar{x}$ and v is limited

x	$R_{100}(x)$	$R_{100}(x)/R_{100}(0.5)$	x	$R_{100}(x)$	$R_{100}(x)/R_{100}(0.5)$
0.45	1.004752335E-195	2.389348988E-005	0.45	-6.502732501E-036	2.472957827E-005
0.46	9.250791602E-195	2.199882370E-004	0.46	-5.945496722E-035	2.261043747E-004
0.47	8.120154981E-194	1.931011595E-003	0.47	-5.182822276E-034	1.971002331E-003
0.48	6.809079657E-193	1.619231628E-002	0.48	-4.316216886E-033	1.641436478E-002
0.49	5.464750018E-192	1.299543626E-001	0.49	-3.440470310E-032	1.308394276E-001
0.50	4.205130100E-191	1.000000000E+000	0.50	-2.629536350E-031	1.000000000E+000
0.51	3.107703452E-190	7.390267075E+000	0.51	-1.930237899E-030	7.340601696E+000
0.52	2.209165994E-189	5.253549691E+001	0.52	-1.362994549E-029	5.183402577E+001
0.53	1.512865076E-188	3.597665329E+002	0.53	-9.271956786E-029	3.526080476E+002
0.54	9.994243076E-188	2.376678685E+003	0.54	-6.084838014E-028	2.314034588E+003
0.55	6.377560790E-187	1.516614383E+004	0.55	-3.857446760E-027	1.466968410E+004

x	$R_{100}(x)$	$R_{100}(x)/R_{100}(2)$	x	$R_{100}(x)$	$R_{100}(x)/R_{100}(2)$
1.80	6.545292773E-135	2.385756179E-005	1.80	-2.164536851E+025	2.561277750E-005
1.84	6.028087247E-134	2.197235005E-004	1.84	-1.964634580E+026	2.324735120E-004
1.88	5.292923199E-133	1.929268051E-003	1.88	-1.700396834E+027	2.012064879E-003
1.92	4.439666981E-132	1.618256555E-002	1.92	-1.406181071E+028	1.663921909E-002
1.96	3.564208076E-131	1.299152190E-001	1.96	-1.113195638E+029	1.317234778E-001
2.00	2.743487717E-130	1.000000000E+000	2.00	-8.451004001E+029	1.000000000E+000
2.04	2.028121883E-129	7.392494857E+000	2.04	-6.162732631E+030	7.292308264E+000
2.08	1.442174205E-128	5.256718288E+001	2.08	-4.323606038E+031	5.116085660E+001
2.12	9.879083708E-128	3.600921429E+002	2.12	-2.922586405E+032	3.458271236E+002
2.16	6.528260340E-127	2.379547866E+003	2.16	-1.906082893E+033	2.255451415E+003
2.20	4.167084197E-126	1.518903902E+004	2.20	-1.200993853E+034	1.421125646E+004

e^x $\qquad\qquad\qquad$ $\frac{1}{1+x}$

Figure 3: Evolution of the remainders $R_n(x)$ of the Taylor expansions $e^x = \sum_{k=0}^{\infty} \frac{x^k}{k!}$ and $\frac{1}{1+x} = \sum_{k=0}^{\infty} (-1)^k x^k$ around x=0.5 (above) and x=2 (below); the index n has the value 100. Theoretically, for unlimited n and $a \geqq 1$ not too big.

$$\frac{R_n(x+\frac{u}{n})}{R_n(x)} = (1+\emptyset)\frac{R_n(ax + \frac{au}{n})}{R_n(ax)} = (1+\emptyset)e^{u/x} \qquad \text{(u limited)}$$

So the (local) deterioration of the approximation does not so much depend on the individual function as well as the distance to the origine. Here we took a=4.

$$R_{\omega-1}{}^{G}(x+\frac{xv}{\omega}) = T_{\omega}{}^{G}(x+\frac{xv}{\omega})A_{\omega}{}^{G}(x+\frac{xv}{\omega})$$

$$= (1+\emptyset)a_{\omega}(x+\frac{xv}{\omega})^{\omega}A_{\omega}{}^{G}(x)$$

$$= (1+\emptyset)a_{\omega}x^{\omega}A_{\omega}{}^{G}(x)(1+\frac{v}{\omega})^{\omega}$$

$$= R_{\omega-1}{}^{G}(x).(1+\emptyset)e^{v}$$

So all these remainders possess the same exponential growth on the x/ω-galaxy of x. This implies that in relative sense the quality of the approximation deteriorates the more rapidly as we are near the origine; in fact the speed of deterioration is nearly reciprocal to the distance to the origine. For appreciable x this is illustrated by the formula

$$R_{\omega-1}(x+\frac{u}{\omega}) = (1+\emptyset)R_{\omega-1}(x)e^{\frac{u}{x}},$$

which is valid for limited u. See Fig. 3. Notice that in the case of the entire functions $\varphi(x)$ mentioned above, the uniform exponential behaviour extends to intervals of unlimited length. Indeed, the formula

$$R_{\omega-1}{}^{\varphi}(x+\frac{xv}{\omega}) = R_{\omega-1}{}^{\varphi}(x)(1+\emptyset)e^{v} \qquad \text{(v limited)}$$

is at least valid for $x/\omega \lnsim 1$. The speed of deterioration of the quality of the approximation continues to be nearly reciprocal to the distance to the origine.

If we add a new term $T_{\omega}{}^{G}(x)$ to the sum $S_{\omega-1}{}^{G}(x)$, what is the gain in length of the interval where the approximation is "acceptable"? Let us state this question in a more precise way. Let $P > 0$ be a precision. Suppose $|R_{\omega-1}{}^{G}(x)| \leq P$ on some interval, say, $[0,D_{\omega-1}]$. Can we know approximately $D_{\omega-1}$, or the distance $d_{\omega} \equiv D_{\omega}-D_{\omega-1}$?. We will consider some situations where we can give in a simple way a fairly general asymptotic answer to this question.

First we consider standard functions $f(x) = \sum_{n=0}^{\infty} c_n x^n$ where $\lim_{n\to\infty} \frac{c_{n+1}}{c_n} = c$, with $0 < c < \infty$. We will solve the external equation (see Definition 4.38)

$$S = \{d \mid |R_{\omega}{}^{f}(x+d)| = (1+\emptyset)|R_{\omega-1}{}^{f}(x)|\}$$

where ω is considered unlimited, and x appreciable, not nearly equal to $|1/c|$, nor to any singularity of f. Now for limited v

$$R_{\omega}{}^{f}(x+\frac{v}{\omega}) = (1+\emptyset)\,\frac{c_{\omega+1}(x+v/\omega)^{\omega+1}}{1-c(x+v/\omega)}$$

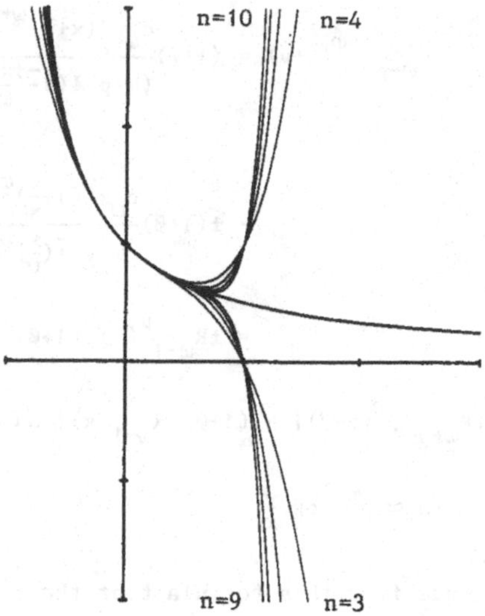

Figure 4: The function $\frac{1}{1+x}$ approached by the partial sums $S_n(x)$ of its Taylor (geometrical) series $\sum_{k=0}^{\infty} (-1)^k x^k$ for n=3 to n=10. Inside the convergence disk the addition of a new term of the series results in an increment of the precision of the approximation, both in length and in depth; outside the convergence disk it is the contrary. These properties are shared by Taylor series of type $\sum_{n=0}^{\infty} c_n x^n$ with $\lim_{n \to \infty} \frac{c_{n+1}}{c_n} = c$.

$$= (1+\emptyset)\frac{c_\omega x^\omega}{1-cx} cx\left(1+\frac{v}{x\omega}\right)^{\omega+1}$$

$$= R_{\omega-1}^{f}(x)(1+\emptyset)cxe^{v/x}$$

Hence $|R_\omega^{f}(x+d)| = (1+\emptyset)|R_{\omega-1}^{f}(x)|$ if and only if

$$d = (1+\emptyset)\frac{x}{\omega} \log \frac{1}{|c|x}$$

We conclude from this formula that there essentially are two cases. If the precision $P > 0$ is very small, such that it is attained on an interval $[0,D_{\omega-1}]$ with $D_{\omega-1}$ inside the convergence disk, the interval where this precision is attained grows by adding a new term ($D_\omega > D_{\omega-1}$). On the contrary if P is large, such that $D_{\omega-1}$ lies outside the convergence disk, the interval where this precision is attained decreases with ω ($D_\omega < D_{\omega-1}$). This of course is no surprise. See Fig. 4.

Secondly we consider standard entire functions $\Psi(x) = \sum_{n=0}^{\infty} \frac{c_n x^n}{n!}$ such that for some $p,q \in \mathbb{N}$ one has $\lim_{n \to \infty} \frac{c_{p(n+1)+q}}{c_{pn+q}} = \pm 1$, while all other coefficients, larger than some (standard) index n_0, are zero. Let ω be unlimited and of the form $\omega = pn+q$. Suppose $0 \lnsim \frac{x}{\omega} \lnsim 1$ and v is limited. Then

$$R_{\omega+p-1}^{\varphi}(x+v) = (1+\emptyset)\frac{c_{\omega+p}(x+v)^{\omega+p}}{(\omega+p)!(1\bar{+}(\frac{x+v}{\omega+p})^{p})}$$

$$= \pm(1+\emptyset)\frac{c_{\omega}x^{\omega}(1+\frac{v}{x})^{\omega+p}(\frac{x}{\omega})^{p}}{\omega!(1\bar{+}(\frac{x}{\omega})^{p})}$$

$$= \pm R_{\omega-1}^{\varphi}(x)\circ(1+\emptyset)e^{\frac{v\omega}{x}}(\frac{x}{\omega})^{p}$$

Hence $|R_{\omega+p-1}^{\varphi}(x+d)| = (1+\emptyset)|R_{\omega-1}(x)|$ if and only if

(1) $$d \simeq p\frac{x}{\omega}\log\frac{\omega}{x}$$

We conclude from this formula that the domains where a certain precision is attained progress nearly linearly with the index. Indeed, for standard k

$$d_{\omega+k} \simeq p.\frac{x}{\omega+k}\log\frac{\omega+k}{x} \simeq p\frac{x}{\omega}\log\frac{\omega}{x} \simeq d_{\omega}$$

The progress of nearly linear progression is strikingly illustrated by Figure 5 and the Figures 17 and 18 of Chapter I.

What is (approximately) the horizontal progression we observe on these figures? We attack this question in the same external way as in section 4 of chapter I.

We interpret "$R_{n-1}(x)$ is observable" by the mathematical expression "$R_{n-1}(x)$ is appreciable"。 Furthermore we treat the question asymptotically: we consider unlimited indices $n = \omega$. If we want to determine those x for which $R_{\omega-1}(x)$ is appreciable we must solve the external equation

$$S = \{x \mid (1+\emptyset)\frac{c_{\omega}}{\omega!}\frac{x^{\omega}}{1\bar{+}(\frac{x}{\omega})^{p}} \in \mathbb{A}\}$$

Using Stirling's formula $\omega! = (1+\emptyset)\omega^{\omega}e^{-\omega}\sqrt{\omega}\sqrt{2\pi}$ and the estimation $c_{\omega} = (1+\emptyset)^{\omega}$ (Lemma 5.10) we easily see that in first approximation $x = (1+\emptyset)\frac{\omega}{e}$. In view of formula (1) this already suffices to know the progression d in first approximation. We have

$$d \simeq p\frac{x}{\omega}\log\frac{\omega}{x} \simeq \frac{p}{e}$$

Numerical data concerning this near-equality are found in chapter I (Fig. 19).

It is interesting to pursue further our investigation of the above external equation. We will derive an explicit asymptotic formula for $R_{\omega-1}(x)$, valid on S.

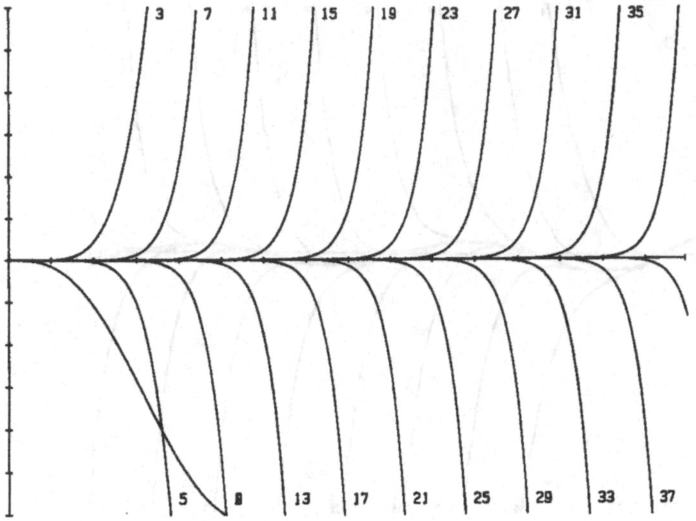

<u>Figure 5</u>: Graphs of the remainders $R_n(x)$ of the Taylor expansion $\sum_{k=0}^{\infty} (-1)^k \frac{x^{2k+1}}{(2k+1)!}$ of sin x for n=1 to n=41 (the number n indicates the degree of the truncated sum). Asymptotically the remainders $R_n(x)$ approach locally -i.e. in the domain where they are appreciable- the exponential curve $u \to e^{eu}$. Furthermore the shift d defined by $|R_{n+2}(x+d)| = |R_n(x)|$ there nearly equals $\frac{2}{e}$. In practice, if we interprete "appreciable" by "observable" these properties are already noticed for $n \geq 5$. It should be stressed that the properties are essentially local: two remainders $R_{n+4}(x)$ and $R_n(x)$ are intersecting.

Because for $x = (1+\emptyset)\frac{\omega}{e}$ the factor $\frac{1}{1\mp(x/\omega)^p}$ is appreciable, we concentrate on the term $T_\omega^{\varphi}(x) = \frac{c_\omega x^\omega}{\omega!}$. We have

$$|T_\omega^{\varphi}(x)| = (1+\emptyset)\frac{e^{\omega \log\frac{xe}{\omega} - \frac{\log \omega}{2} + \log |c_\omega|}}{\sqrt{2\pi}}$$

Using the approximation $\log(1+\alpha) = \alpha(1+\emptyset)$ we easily see that the exponent is limited, and a fortiori $T_\omega^{\varphi}(x)$ is appreciable if x is of the form

$$x = \omega' + u \qquad\qquad u \text{ limited}$$

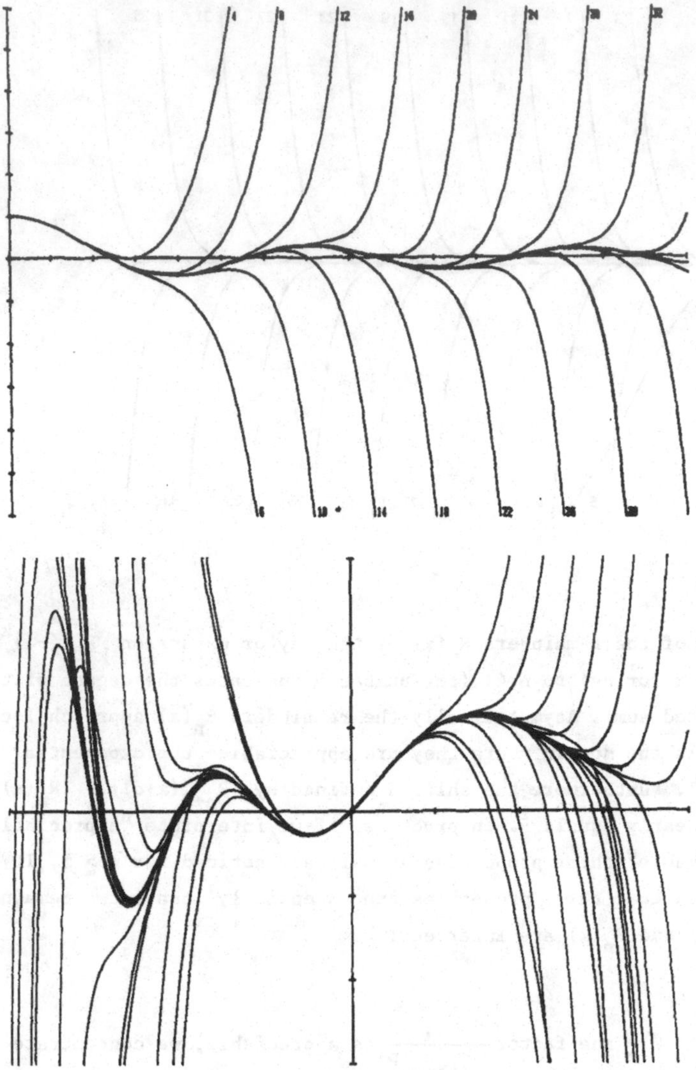

<u>Figure 6</u>: Regular and irregular(?) behaviour of Taylor polynomials of entire functions (above). The Besselfunction $J_0(x) = \sum_{k=0}^{\infty}(-1)^k \frac{x^{2k}}{k!k!}$ is approached by its Taylor polynomials in the same regular way as the sine function. It is not difficult to adapt the external argument applied to the series $\sum_{k=0}^{\infty} \frac{c_k x^k}{k!}$ to this case (Below). It seems that the Taylor polynomials of $\frac{1}{\Gamma(x)}$ behave irregularly. Here they are shown for n=3 to n=21.

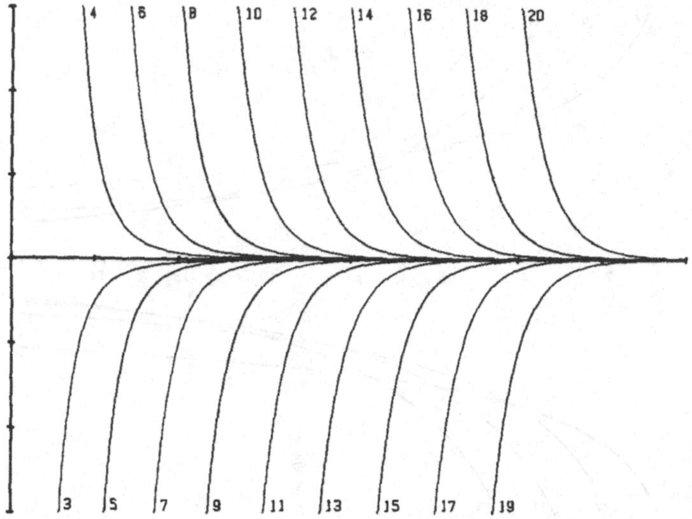

Figure 7: The remainders $R_n(x)$ of the divergent asymptotic expansion $\sum_{k=0} (-1)^k \frac{(\frac{1}{2})^k}{x^{k+1}}$ of the Laplace transform $F(x) = {}_0\int^{\infty} \frac{e^{-tx}}{\sqrt{1+t}}dt$ *for* n=3 to 20. The aspect of the configuration is similar to that of remainders of convergent Taylor series of type $\sum_{k=0}^{\infty} c_k \frac{x^k}{k!}$ with $\lim_{k \to \infty} \frac{c_{k+1}}{c_k} = -1$. The difference is that here they behave on the domains where they are observed as an exponential function with negative exponent ($u \to e^{-eu}$, in absolute value the exponent is the same). Furthermore they are "receding" towards infinity with a nearly constant shift ($|R_{n-1}(x)| = |R_n(x+d)|$ for d equal to about 1/e, this shift is the same as in the convergent case).

where

$$\omega' = \frac{\omega}{e} + \frac{\log \omega}{2e} - \frac{\log|c_\omega|}{e}$$

Then for all limited u

$$|R_{\omega-1}(\omega'+u)| \simeq \frac{1}{\sqrt{2\pi}} \frac{e^{eu}}{1\mp(1/e)^p}$$

This expression is indeed nearly invariant as a function of ω, which confirms our visual observations.

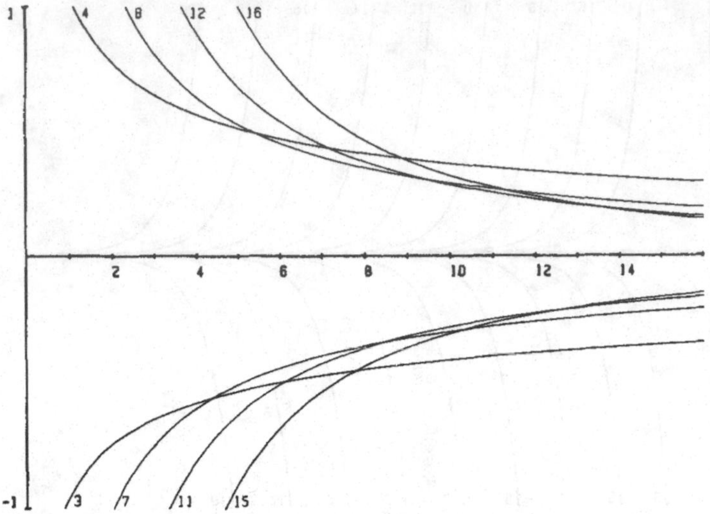

<u>Figure 8</u>: Remainders $R_n(x)$ of the asymptotic expansion $\sum_{k=0}(-1)^k \frac{(\frac{1}{2})^k}{x^{k+1}}$ of $\int_0^\infty \frac{e^{-tx}}{\sqrt{1+t}}dt$ (see Fig. 7) now seen under the Benoit micrococoscope. That is to say, we plotted

$$\begin{cases} R_n(x)^\varepsilon & R_n(x) > 0 \\ -(-R_n(x))^\varepsilon & R_n(x) < 0 \end{cases}$$

where we chose $\varepsilon = \frac{1}{10}$ and n=3,4,7,8,11,12,15,16. Notice that this exponential micrococoscope "desingularizes" Fig. 7, enabling us to observe that the curves representing the remainders are intersecting. The drawing confirms the asymptotic results of the text: Suppose at a point x we obtained a precision P by summing to the index n, and wish to add a new term. If the argument is well larger than the index, the precision at x increases, together with the domain where the former precision P is attained: we gain both in depth and in width. On the other hand, if the index is well larger than the argument, the effect is desastrous: at the point x we loose the precision P, and the interval where the precision P is attained shrinks.

Finally, let us study the horizontal progression of the remainders in the case of a class of divergent expansions.

We consider Laplace transforms

$$L(x) = \int_0^\infty e^{-tx} f(t)dt$$

where $f(t) = \sum_{n=0}^{\infty} c_n t^n$, with $\lim_{n \to \infty} \frac{c_{n+1}}{c_n} = -1$. Then L(x) has the divergent asymptotic expansion at infinity

$$L(x) \sim \sum_{n=0} \frac{c_n n!}{x^{n+1}}$$

As a consequence of Theorem 2.16 the following asymptotic expression for the remainder holds:

$$R_{\omega-1}{}^{L}(x) = (1+\emptyset)\frac{T_\omega{}^{L}(x)}{1+\omega/x}$$

If ω is unlimited this formula is valid at least for those unlimited x such that ω/x is limited and not nearly equal to 1. Again these remainders behave locally as an exponential function, but now the exponent is negative. In particular, we find in the same manner as in the convergent case

$$R_{\omega-1}{}^{L}(x+v) = (1+\emptyset)R_{\omega-1}{}^{L}(x)e^{-\frac{\omega}{x}v}$$

We may also determine the horizontal progression in the same manner as above. We find that $|R_\omega{}^{L}(x+d)| = (1+\emptyset)|R_{\omega-1}{}^{L}(x)|$ if and only if

$$d \simeq -\frac{x}{\omega}\log\frac{x}{\omega}$$

We conclude that if $x/\omega \gtrsim 1$, the distance d_ω is negative; notice that the domain where a given precision is attained then increases (we already knew that for fixed x much larger than the index the remainder $R_\omega(z)$ is smaller than $R_{\omega-1}(x)$). On the other hand if $x/\omega \lesssim 1$, the distance d_ω is positive. Then the domain where some precision is attained decreases (again we knew that for fixed x much smaller than the index the remainder $R_\omega(x)$ is larger than $R_{\omega-1}(x)$). See Figures 7 and 8.

P A R T III

CHAPTER IV. EXTERNAL SETS.

In classical Zermelo-Fraenkel set theory with the axiom of choice (ZFC) in principle all sets are defined using the only non-logical symbol $\in$. In the nonstandard extension IST there is also the possibility to define collections with the non-logical symbol st. Those collections which fall outside the range of ZFC are called external sets. External sets are often easily recognized: mostly some elementary classical property fails to hold. For instance, the set $\underline{\mathbb{N}} = \{\text{st } n \mid n \in \mathbb{N}\}$ satisfies the premisses of induction ($0 \in \underline{\mathbb{N}}$ and $n \to \underline{\mathbb{N}} \Rightarrow n+1 \in \underline{\mathbb{N}}$), but not its conclusion, for $\underline{\mathbb{N}} \neq \mathbb{N}$. So $\underline{\mathbb{N}}$ is external. Likewise the set of infinitesimal real numbers hal(0) must be external, for it constitutes a bounded subset of $\mathbb{R}$ without lower upper bound.

In this book external sets play an important part. Firstly we used them to model qualitative transitions and vague boundaries (Chapters I and III). Secondly, they act as orders of magnitude; this has also a practical meaning, for recognition of the order of magnitude of a phenomenon yields an appropriate change of local variables to study the phenomenon (Chapters I, III and V). Thirdly, they act as domains of asymptotic behaviour, and as domains of validity of reasonings. Often those domains are not the same and here lies a fourth range of application of external sets: they induce permanence principles, which automatically extend results derived directly on some domain to neighbouring regions.

This chapter contains a detailed study of external sets whose elements are internal. First we discuss permanence (section 1). In section 2 we treat the simplest and most widely used types of external sets, namely those defined with one external quantifier of the form ($\forall$ st $n \in \mathbb{N}$) or ($\exists$ st $n \in \mathbb{N}$): "halos" and "galaxies". in section 3 we discuss external sets in a more general setting. It turns out that there are only three main types. In sections 4 and 5 we treat the class which has been the most useful to our work in asymptotics: the convex external subsets of ordered sets, and in particular those which are a "halo" or a "galaxy".

From a didactical point of view, it should be mentioned that not all material presented here has been relevant to nonstandard asymptotics. Notably the contents of section 3 should be seen as generalizations of results about halo's and galaxies. But even in this "logical digression" the results tend to be rather concrete, and the proofs are not in the least technical; furthermore, sometimes such a theoretical result has a quite nice application. For instance, the existence of a universal "sum" to shadow expansions (Chapter VI) is a consequence of the principle of the universal index.

It should also be pointed out that actually there does not exist an axiomatic system (as an extension of IST) which is entirely satisfactory for the manipulation of external sets. So sometimes we will proceed "naively".

1. What is Permanence?

It happens sometimes in classical mathematics that a property is assumed, or proved, on a certain domain, and that afterwards it is remarked that the character of the property and the nature of the domain are incompatible. So actually the property must be valid on a larger domain. The following example, taken from topology, is an illustration of this phenomenon. Let $f: \mathbb{R} \to \mathbb{R}$ be a continuous function such that $f(x) = 0$ on $]0,1[$. Then it follows from the difference between open and closed sets that the property "$f(x) = 0$" is in fact valid on a set which is larger than $]0,1[$; indeed, $\{x \mid f(x) = 0\}$ is closed, and $]0,1[$ is open. Now arguments which take into account the difference between two kinds of sets do not often occur in classical mathematics, one of the more frequently used being the possibility to define a real continuous function on the mere rationals. However, arguments of this kind are frequently used in nonstandard analysis. Recall that nonstandard analysis is intrinsically based upon the distinction of types of sets: standard versus nonstandard, internal versus external. As will be shown, besides these necessary distinctions, other incompatible classes can be defined.
In nonstandard analysis, statements which affirm that the validity of a property exceeds the domain where it was established in direct way are called continuity principles (Stroyan-Luxemburg [36]) or permanence principles (Lightstone-Robinson [34]). But following Lutz-Goze [4] and M. Diener [23] we slightly shift our attention towards what we consider as the background of those permanence principles, namely statements which express the difference between two classes of sets.

In the remaining part of this section we briefly discuss permanence resulting from the difference between standard, internal and external sets. In section 2 and 3, we will define mutually exclusive subclasses of the class of external sets; the resulting permanence principles will be discussed more thoroughly, as they are more important for our work in asymptotics.

First, an infite standard set is larger than the (external) set of all its standard elements. Here the transfer principle

$$\forall^{st} t\ (\forall^{st} x\ A(x,t) \to \forall x\ A(x,t))$$

(A a ZFC-formula) acts as a permanence principle: in order to hold for all elements of a standard set, a standard property needs only to be proved for all standard elements of the set. In this context, let us make the following historical remark. As already said in the introduction, the theory of neutrices of Van der Corput bears some resemblance to nonstandard analysis. In [41, p. 33] he calls "permanence principle" a statement which a nonstandard analyst would recognize as a transfer principle.

Secondly, many permanence results used in nonstandard analysis are based upon the self evident statement "no external set is internal". Many of these applications can be reduced to the following proposition. The proposition is powerful because it is a statement about permanence in the sense of an order relation.

Proposition: Let P be an internal property such that P(n) for all st $n \in \mathbb{N}$. Then there exists some unlimited $\omega \in \mathbb{N}$ such that P(n) for all $n \leq \omega$.

Proof: Put $I = \{k \in \mathbb{N} \mid (\forall n \leq k)P(n)\}$. Then $\underline{\mathbb{N}} \subset I$. Now $\underline{\mathbb{N}}$ is external and I is internal, so $\underline{\mathbb{N}} \neq I$. Hence $\underline{\mathbb{N}} \subsetneq I$. Any element of $I \setminus \underline{\mathbb{N}}$ will do.

For instance, let $(a_n)_{n\in\mathbb{N}}$ be an internal sequence such that $|a_n| \leq 1/n$ for all st n (such as given by the law $a_n = n\varepsilon$, or $a_n = \varepsilon^{1/n}$, where $\varepsilon > 0$, $\varepsilon \simeq 0$). Then we may conclude automatically, and without further calculation, that $|a_n| \leq 1/n$ up to some unlimited index ω. This is especially interesting when the inequality $|a_n| \leq 1/n$ is obtained using in essential way the fact that n is limited (in the above examples it is tempting to use the argument "n limited $\Rightarrow |a_n| \simeq 0 \Rightarrow |a_n| \leq 1/n$"). It may very well be that this inequality is still needed up to some unlimited ω, and that it is more difficult to prove the inequality directly for those larger indices.

The above proposition is called the "Cauchy principle" by Stroyan and Luxemburg. Cauchy distinguishes among the reals numbers which are "specified", like 1,2,3,1/10,... and "variables" like the infinitesimals. It is clear to him that being "specified" is a different kind of property than, say, being "between 0 and 1", for the refuses to quantify over the specified numbers [31].

We propose to extend the name "Cauchy principle" to the statement "no external set is internal". We conclude this section with four typical applications.

1) Let I be an internal set and $f: I \to \mathbb{R}$ be an internal function such that $f(x) \simeq 0$ for all $x \in I$. Then $\sup_{x \in I} f(x) \simeq 0$. Indeed, $\{r \in \mathbb{R} \mid r \gtrsim 0\} \subset \{M \in \mathbb{R} \mid (\forall x \in I)f(x) \leq M\}$. The first set is external and the second set is internal. Hence the inclusion is strict.

2) Let $S \subset \mathbb{N}$ be standard. Suppose S contains an unlimited element ω. Let ω' be an arbitrary unlimited number. Then S contains an unlimited element $\nu \leq \omega'$. Indeed, $S \neq \emptyset$, and $(\forall^{st} n \in S)(\exists m \in S)(n < m)$, simply take $m = \omega$. By transfer, $(\forall n \in S)(\exists m \in S)(n < m)$. So S is infinite. Hence $\{st\ s \mid s \in S\}$ is external. This external set is included in the internal set $\{s \in S \mid s < \omega'\}$. By the Cauchy principle, the inclusion is strict.

3) Let $(u_n)_{n\in\mathbb{N}}$ be a standard sequence such that $u_n \simeq r$ for all unlimited n up to some unlimited ω. Then $u_n \simeq r$ for all unlimited n. Indeed, first suppose r is unlimited. By the Cauchy principle $\{n \leq \omega \mid u_n \text{ unlimited}\} \subsetneq \{n \leq \omega \mid |u_n - r| \leq 1\}$. So

there is some st $\bar{n}$ such that $u_{\bar{n}}$ is nonstandard, a contradiction. Hence r is limited. Let $s = {}^o r$ be its shadow. Suppose there is $n > \omega$ such that $u_n \not\simeq r$, or equivalently, $u_n \not\simeq s$. Then there exists standard d such that $|u_n - s| > d$. By 2) there exists some unlimited $\nu \leq \omega$ such that $|u_\nu - s| > d$, a contradiction. So $u_n \simeq r$ for all unlimited n (Actually, $\lim_{n \to \infty} u_n = s$). See also [35].

2. Halos and Galaxies.

The material of this section may be seen as the set theoretical implications of Robinson's sequential lemma. Much of this section is based on joint work with M. Diener, who was the first to notice the set theoretical meaning of this well-known lemma.

Definition of (pre)galaxies and (pre)halos.

Definition 4.1: A set G is called a galaxy if (i) G is external and (ii) there is an internal sequence $(A_n)_{n \in \mathbb{N}}$ of internal sets such that $G = \bigcup_{n \in \mathbb{N}} A_n$. A set H is called a halo if (i) H is external and (ii) there is an internal sequence $(B_n)_{n \in \mathbb{N}}$ of internal sets such that $H = \bigcap_{n \in \mathbb{N}} B_n$. A set which satisfies the second requirement for a galaxy but not necessarily the first is called pregalaxy. A set which satisfies the second requirement for a halo, but not necessarily the first is called a prehalo.

Comments. 1. Strictly spoken, we could have said "internal sequence $(A_n)_{n \in \mathbb{N}}$" instead of "internal sequence $(A_n)_{n \in \mathbb{N}}$ of internal sets" for, of course, if $(A_n)_{n \in \mathbb{N}}$ is an internal set, so are all A_n.

Another possible weakening of the above definition is "sequence $(A_n)_{n \in {}^\sigma\mathbb{N}}$" instead of "internal sequence $(A_n)_{n \in \mathbb{N}}$". Indeed, an external sequence of internal sets $(A_n)_{n \in {}^\sigma\mathbb{N}}$ may be extended to an internal sequence $(A_n)_{n \in \mathbb{N}}$ on behalf of the extension principle [6].

2. Here are some examples. Most of them are already met in Part I.

1). Every internal set is both a pregalaxy and a prehalo.

2). The set $\pounds$ of the limited real numbers is a galaxy. It is called the principal galaxy after Robinson. In fact, Robinson's terminology was borrowed to serve a wider purpose.

For all $\alpha \neq 0$ and a in $\mathbb{R}$ the α-galaxy of a (notation: α-gal(a)) is defined by

$$\alpha\text{-gal}(a) = \{x \in \mathbb{R} \mid \tfrac{x-a}{\alpha} \text{ limited}\}$$

The α-galaxy of 0 may be seen as all real numbers "of order α". Other galaxies are $\underline{\mathbb{N}}$ and the set $\mathbb{A}$ of all appreciable real numbers:

$$\mathbb{A} = \{x \in \mathbb{R} \mid x \text{ limited, } x \not\simeq 0\}$$

3) Let $a \in \mathbb{R}$. The set of all real numbers x such that x-a is infinitesimal is a halo, called the halo of a. It will be written hal(a)). Likewise we define for all $\alpha \not\simeq 0$ and $a \in \mathbb{R}$ the α-halo of a (notation: α-hal(a)) by

$$\alpha\text{-hal}(a) = \{x \in \mathbb{R} \mid \frac{x-a}{\alpha} \simeq 0\}$$

Other halo's are $\widetilde{\mathbb{N}}$, the set of all unlimited natural numbers, and the set of the unlimited real numbers.

4) Let $A \subset \mathbb{R}^n$ be internal, where st n. The halo of A (notation: hal(A)) is defined by

$$\text{hal}(A) = \{x \in \mathbb{R}^n \mid |x-a| \simeq 0 \text{ for some } a \in A\}$$

5) The set of all internal real functions f such that $f(x) \simeq 0$ for all limited x is a prehalo for

$$(\forall x \in \mathbb{L})(f(x) \simeq 0) \Rightarrow (\forall n \in \underline{\mathbb{N}})(f[-n,n] \subset [-1/n,1/n])$$

6) External sets which are neither a galaxy nor a halo are treated in section 3. For instance, it will be shown that the set of all standard elements of an uncountable standard set is not a galaxy and not a halo.

3. The examples given above, and those already encountered in part I of this book show that many "natural" external sets belong to the class of the halo's, or to the class of the galaxies. In a way, galaxies and halo's are the simplest external sets that one may come across. They are defined with one reference to the most elementary external set, $\underline{\mathbb{N}}$, using one quantifier, which is existential in the case of galaxies and universal in the case of halo's.

4. Let T be a standard topological space. Let X be a standard index set and $(A_x)_{x \in X}$ be a standard family of open sets. Then $\bigcup_{x \in X} A_x$ will be called a union monad or a topological galaxy and $\bigcap_{x \in X} A_x$ will be called an intersection monad or a topological halo. Some topological galaxies are galaxies in our sense, like $\mathbb{L}$ or $\mathbb{A}^+$, and some topological halos are halos in our sense , like hal(a) for every st $a \in \mathbb{R}$. The monad of a standard point a of T is the intersection of all open neighbourhoods of a

If T has a countable neighbourhood basis, then this monad is a prehalo.

Topological spaces are discussed from a nonstandard point of view notably in [36].

The following monotony-lemma allows us to work in a more comfortable way with the notions of pregalaxy or prehalo.

<u>Lemma 4.2</u>: Let G be a pregalaxy and H be a prehalo. Then there exists an internal increasing sequence $(A_n)_{n\in\mathbb{N}}$ and an internal decreasing sequence $(B_n)_{n\in\mathbb{N}}$ of internal sets such that

$$G = \bigcup_{n\in\underline{\mathbb{N}}} A_n = \bigcap_{n\in\overline{\mathbb{N}}} A_n \qquad\qquad H = \bigcap_{n\in\underline{\mathbb{N}}} B_n = \bigcup_{n\in\overline{\mathbb{N}}} B_n$$

<u>Proof</u>: The statement about the halo being the immediate consequence of the statement about the galaxy, we prove the lemma only for galaxies. Let G be a galaxy and $(I_n)_{n\in\mathbb{N}}$ be an internal sequence of internal sets such that $G = \bigcup_{n\in\underline{\mathbb{N}}} I_n$. Define for each $n \in \mathbb{N}$ the set A_n by $A_n = \bigcup_{k\leq n} I_k$.

In order to prove that $\bigcup_{n\in\underline{\mathbb{N}}} A_n = \bigcap_{n\in\overline{\mathbb{N}}} A_n$ we remark first that the monotony of the sequence $(A_n)_{n\in\mathbb{N}}$ ensures that $\bigcup_{n\in\underline{\mathbb{N}}} A_n \subset \bigcap_{n\in\overline{\mathbb{N}}} A_n$. Conversely, suppose $x \in A_n$ for all $n \in \overline{\mathbb{N}}$. Then by the Cauchy principle there is $m \in \underline{\mathbb{N}}$ such that $x \in A_m$. So $\bigcap_{n\in\overline{\mathbb{N}}} A_n \subset \bigcup_{n\in\underline{\mathbb{N}}} A_n$. Hence $\bigcup_{n\in\underline{\mathbb{N}}} A_n = \bigcap_{n\in\overline{\mathbb{N}}} A_n$.

<u>Elementary operations on pregalaxies and prehalos.</u>

Here we describe the behaviour of the notion of pregalaxy and prehalo under elementary set theoretic operations.

<u>Proposition 4.3</u>: The complement of a pregalaxy in an internal set is a prehalo.

<u>Proposition 4.4</u>: The image and inverse image of a pregalaxy under an internal mapping are pregalaxies. The image and inverse image of a prehalo under an internal mapping are prehalos.

<u>Proposition 4.5</u>: If $(G_n)_{n\in\underline{\mathbb{N}}}$ is a sequence of pregalaxies, then $\bigcup_{n\in\underline{\mathbb{N}}} G_n$ is a pregalaxy. If $(H_n)_{n\in\underline{\mathbb{N}}}$ is a sequence of prehalos, then $\bigcap_{n\in\underline{\mathbb{N}}} H_n$ is a prehalo.

<u>Proof</u>: We only prove the statement about the pregalaxies. For all $n \in \mathbb{N}$, let $(A_{nk})_{k\in\mathbb{N}}$ be an internal and increasing sequence such that $G_n = \bigcup_{k\in\underline{\mathbb{N}}} A_{nk}$. We define for every $m \in \underline{\mathbb{N}}$ the internal set B_m by $B_m = \bigcup_{n,k\leq m} A_{nk}$. Then $\bigcup_{n\in\underline{\mathbb{N}}} G_n = \bigcup_{m\in\underline{\mathbb{N}}} B_m$. Hence $\bigcup_{n\in\underline{\mathbb{N}}} G_n$ is a pregalaxy.

Proposition 4.6: The set of all internal subsets of a pregalaxy is a pregalaxy. The set of all internal subsets of a prehalo is a prehalo.

Proof: We only prove the statement about the pregalaxies. Let G be a pregalaxy and $(A_n)_{n\in\mathbb{N}}$ be an internal increasing sequence such that $G = \bigcup_{n\in\mathbb{N}} A_n$. Let X be an internal set such that $G \subset X$. Put $S_n = \{S \subset X \mid S \subset A_n\}$. Then S_n is internal and $\{S \subset G \mid S \text{ internal}\} = \bigcup_{n\in\mathbb{N}} S_n$.

Proposition 4.7: Let $G \subset X$ be a pregalaxy and $H \subset Y$ be a prehalo, where X and Y are internal sets. Then the set of all internal mappings $f: X \to Y$ such that $f(G) \subset H$ is a prehalo and the set of all internal mappings $f: Y \to X$ such that $f(H) \subset G$ is a pregalaxy.

Proof: (i) Let $(A_n)_{n\in\mathbb{N}}$ be an internal increasing sequence such that $G = \bigcup_{n\in\mathbb{N}} A_n$ and let $(B_n)_{n\in\mathbb{N}}$ be an internal decreasing sequence such that $H = \bigcap_{n\in\mathbb{N}} B_n$. Now

$$f(G) \subset H \Leftrightarrow (\forall n \in \mathbb{N})(\forall m \in \mathbb{N})(f(A_n) \subset B_m)$$

So, by proposition 4.5, the set of all internal mappings $f: X \to Y$ such that $f(G) \subset H$ is a prehalo.

(ii) The second statement will be proved later on using the Fehrele principle (application 3).

As a consequence of proposition 4.4, every inverse image $f^{-1}(\mathbb{N})$, where f is some internal mapping, is a pregalaxy, and in the same way every inverse image $f^{-1}(\text{hal}(0))$ is a prehalo. In fact pregalaxies and prehalos may be characterized that way, as is shown by the next proposition.

Proposition 4.8: (i) A subset G of an internal set X is a pregalaxy if and only if G is the inverse image of $\mathbb{N}$ under an internal mapping from X into $\mathbb{N}$.

(ii) A subset H of an internal set X is a prehalo if and only if H is the inverse image of hal(0) under an internal mapping from X into $\mathbb{R}$.

Proof: (i) Let $G \subset X$ be a pregalaxy and let $(A_n)_{n\in\mathbb{N}}$ be an internal increasing sequence of internal sets such that $G = \bigcup_{n\in\mathbb{N}} A_n$. We may assume that $\bigcup_{n\in\mathbb{N}} A_n = X$. We define the internal mapping $p: X \to \mathbb{N}$ by

$$p(x) = \min\{n \in \mathbb{N} \mid x \in A_n\}$$

Clearly, $G = p^{-1}(\mathbb{N})$. Conversely, $p^{-1}(\mathbb{N})$ is a pregalaxy for every internal mapping $p: X \to \mathbb{N}$ by proposition 4.4.

(ii) Let $H \subset X$ be a prehalo and let $(B_n)_{n\in\mathbb{N}}$ be an internal decreasing sequence of internal sets such that $H = \bigcap_{n\in\underline{\mathbb{N}}} B_n$. We may assume that $B_0 = X$. We define the internal mapping $p: X \to \mathbb{R}$ by

$$p(x) = \inf\{1/n \mid n \in \mathbb{N} \text{ and } x \in B_n\}$$

Clearly $H = p^{-1}(\text{hal}(0))$. Conversely, $p^{-1}(\text{hal}(0))$ is a pregalaxy for every internal mapping $p: X \to \mathbb{R}$ by proposition 4.4.

Pregalaxies and prehalos were introduced by M. Diener in [23], using the above characterization as a definition. The characterization is very useful for identifying pregalaxies and prehalos in practice. Indeed, if the word "limited" occurs in the definition of a set, this set will very often be a pregalaxy, and if the word "infinitesimal" occurs in the definition of a set, this set will very often be a prehalo.

However, in a theoretical context, the sequence characterizations are easier to work with.

Distinguishing internal and external sets.

It is often needed to distinguish external sets from internal sets, for instance in applying permanence. In next theorem and its corollary, methods are presented to decide whether a pregalaxy is a galaxy or a prehalo is a halo.

Theorem 4.9: (i) A set G is a galaxy if and only if there exists a strictly increasing sequence of internal sets $(A_n)_{n\in\underline{\mathbb{N}}}$ such that $G = \bigcup_{n\in\underline{\mathbb{N}}} A_n$.

(ii) A set H is a halo if and only if there exists a strictly decreasing sequence of internal sets $(B_n)_{n\in\underline{\mathbb{N}}}$ such that $H = \bigcap_{n\in\underline{\mathbb{N}}} B_n$.

Proof: We only prove (i). Let G be a galaxy and let $(B_k)_{k\in\mathbb{N}}$ be an internal increasing sequence of internal sets such that $G = \bigcup_{k\in\underline{\mathbb{N}}} B_k$. We define a strictly increasing subsequence $(A_n)_{n\in\underline{\mathbb{N}}}$ of the sequence $(B_k)_{k\in\underline{\mathbb{N}}}$ as follows. We remark first that $B_k \subsetneq G$ for all standard $\overline{k}$, for G is external. So there exists for all standard k a standard natural number $p > k$ such that $B_k \subsetneq B_p$. Using external induction we find a strictly increasing sequence of standard natural numbers $(k_n)_{n\in\underline{\mathbb{N}}}$ such that $B_{k_n} \subsetneq B_{k_{n+1}}$ on $\underline{\mathbb{N}}$. Putting $A_n = B_{k_n}$ we obtain a strictly increasing sequence $(A_n)_{n\in\underline{\mathbb{N}}}$ such that $G = \bigcup_{n\in\underline{\mathbb{N}}} A_n$.

Conversely, let $(A_n)_{n\in\underline{\mathbb{N}}}$ be a strictly increasing sequence of internal sets. Put $G = \bigcup_{n\in\underline{\mathbb{N}}} A_n$. By the principle of extension there exists an internal prolongation $(A_n)_{n\in\mathbb{N}}$ of this sequence, that we may assume increasing in virtue of the Cauchy principle. So G is a pregalaxy. Suppose G is internal. Now $A_n \subsetneq G$ for all $n \in \underline{\mathbb{N}}$, so by the Cauchy principle there exists $\nu \in \overline{\mathbb{N}}$ such that $A_\nu \subsetneqq G$. Hence $\bigcup_{n\in\underline{\mathbb{N}}} A_n \subsetneqq G$, a contradiction. Hence G is external and thus G is a galaxy.

Corollary 4.10: Let X and Y be two internal sets, let $f: X \to Y$ be an internal mapping, let $G_1 \subset X$ and $G_2 \subset Y$ be two galaxies and $H_1 \subset X$ and $H_2 \subset Y$ be two halos. Then

(i) if $f \restriction G_1$ is one-to-one, then $f(G_1)$ is a galaxy, if $f \restriction H_1$ is one-to-one, then $f(H_1)$ is a halo.

(ii) If f is onto G_2, then $f^{-1}(G_2)$ is a galaxy, if f is onto H_2, then $f^{-1}(H_2)$ is a halo.

The Fehrele principle.

No galaxy is a halo. This follows from next theorem, which states that a galaxy and a halo are always "separated" by an internal set.

Theorem 4.11: If G is a galaxy and H is a halo such that $G \subset H$, then there exists an internal set I such that $G \subset I \subset H$.

Proof: Let $(A_n)_{n\in\underline{\mathbb{N}}}$ be an internal increasing sequence of internal sets such that $G = \bigcup_{n\in\underline{\mathbb{N}}} A_n$, and let $(B_n)_{n\in\mathbb{N}}$ be an internal sequence of internal sets such that $H = \bigcap_{n\in\underline{\mathbb{N}}} B_n$. As $A_n \subset B_n$ for all $n \in \underline{\mathbb{N}}$ there exists by the Cauchy principle an unlimited ν such that $A_n \subset B_n$ for all natural numbers n up to ν. Then

$$G = \bigcup_{n\in\underline{\mathbb{N}}} A_n \subset \bigcup_{n\leqq\nu} A_n = A_\nu \subset B_\nu = \bigcap_{n\leqq\nu} B_n \subset \bigcap_{n\in\underline{\mathbb{N}}} B_n = H$$

Putting, for instance, $I = A_\nu$ we obtain that $G \subset I \subset H$.

Theorem 4.12: (The Fehrele[1] principle). No halo is a galaxy.

Proof: Let G be a galaxy and H be a halo. Assume $G \subset H$. By Theorem 4.11, we may let I be an internal set such that $G \subset I \subset H$. By the Cauchy principle $G \subsetneqq I \subsetneqq H$. Hence $G \neq H$.

The Fehrele principle shows among other things, that properties based on the notion "limited" and properties based on the notion "infinitesimal" are incompatible. Confrontation of these notions leads to many permanence results. We adopt the following convention. Suppose a pregalaxy G is set against a prehalo H, and one of these sets is proved to be external. Then the conclusion "$G \neq H$ by the Cauchy principle, if one set is external, or by the Fehrele principle, if both sets are

---

(1) See page 20.

external" will be shortened to "$G \neq H$ by the Fehrele principle".

Some applications of the Fehrele principle.

1. Robinson's sequential lemma [35]. If $(a_n)_{n\in\mathbb{N}}$ is an internal sequence of real numbers such that $a_n \simeq 0$ for all limited $n \in \mathbb{N}$, then there exists an unlimited $\omega \in \mathbb{N}$ such that $a_n \simeq 0$ for all $n \leq \omega$.

Proof: One has $\underline{\mathbb{N}} \subset \{k \in \mathbb{N} \mid (\forall n \leq k)(a_n \simeq 0)\}$. The first set is a galaxy while the second is a prehalo. By the Fehrele principle, the inclusion is strict. Hence $a_n \simeq 0$ up to some unlimited $\omega \in \mathbb{N}$.

The lemma was originally proved by Robinson applying the Cauchy principle to the inclusion $\underline{\mathbb{N}} \subset \{k \in \mathbb{N} \mid (\forall n \leq k)(|a_n| \leq 1/n)\}$. The lemma is very useful, being a permanence property in the sense of an order relation.

2. Lemma of dominated approximation. Let f, g be two real internal (measurable or Riemann-integrable) functions, such that $f(x) \simeq g(x)$ for all limited $x \in \mathbb{R}$. Let h be a standard integrable function such that $|f(x)|,|g(x)| \leq h(x)$ for all $x \in \mathbb{R}$. Then $\int_{-\infty}^{\infty} f(x)dx \simeq \int_{-\infty}^{\infty} g(x)dx$.

This lemma has been most useful to me in the study of asymptotic approximations. It institutionalizes a characteristic manner of applying permanence. Indeed, it follows directly from the assumptions that $\int_{-a}^{a} f(x)dx \simeq \int_{-a}^{a} g(x)dx$ as long as a is limited and $\int_{|x|\geq\omega} f(x)dx \simeq \int_{|x|\geq\omega} g(x)dx$ as long as ω is unlimited. The "gap" is filled by permanence, for the prehalo $\{a \mid \int_a^a f(x)dx \simeq \int_a^a g(x)dx\}$ exceeds the galaxy $\mathfrak{L}$.

3. (Second half of proposition 4.7). If X and Y are internal sets and $G \subset X$ is a pregalaxy and $H \subset Y$ is a prehalo, then the set S of all internal mappings $f: Y \to X$ such that $f(H) \subset G$ is a pregalaxy.

Proof: Let $(A_n)_{n\in\mathbb{N}}$ be an internal increasing sequence of internal sets such that $G = \bigcup_{n\in\underline{\mathbb{N}}} A_n$, and let $(B_n)_{n\in\mathbb{N}}$ be an internal decreasing sequence of internal sets such that $H = \bigcap_{n\in\underline{\mathbb{N}}} B_n$. We are going to prove that $S = \bigcup_{m\in\underline{\mathbb{N}}} \bigcup_{n\in\underline{\mathbb{N}}} A_n^{B_m}$. Clearly, if f is an internal function such that $f(B_m) \subset A_n$ for some $m,n \in \underline{\mathbb{N}}$, then $f(H) \subset G$. Conversely, let f be an internal function such that $f(H) \subset G$. Put $m_n = \min\{m \mid f(B_m) \subset A_n\}$ for every $n \in \mathbb{N}$. Then $(m_n)_{n\in\mathbb{N}}$ is an internal sequence of natural numbers. Suppose $m_n \in \overline{\mathbb{N}}$ for all $n \in \underline{\mathbb{N}}$. Then there exists by the Fehrele principle $\nu \in \overline{\mathbb{N}}$ such that $m_\nu \in \overline{\mathbb{N}}$. So there exists $x \in B_{m_\nu - 1}$ such that $f(x) \notin A_\nu$. This is a contradiction, for $x \in H$ and $f(x) \notin G$. So there is $n \in \underline{\mathbb{N}}$ such that $m_n \in \underline{\mathbb{N}}$, which implies that $f(B_{m_n}) \subset A_n$. So $S = \bigcup_{m\in\underline{\mathbb{N}}} \bigcup_{n\in\underline{\mathbb{N}}} A_n^{B_m}$. Hence S is a pregalaxy by proposition 4.5.

4. <u>No trespassing!</u> Consider the following quadrilation of $\mathbb{R}^2$.

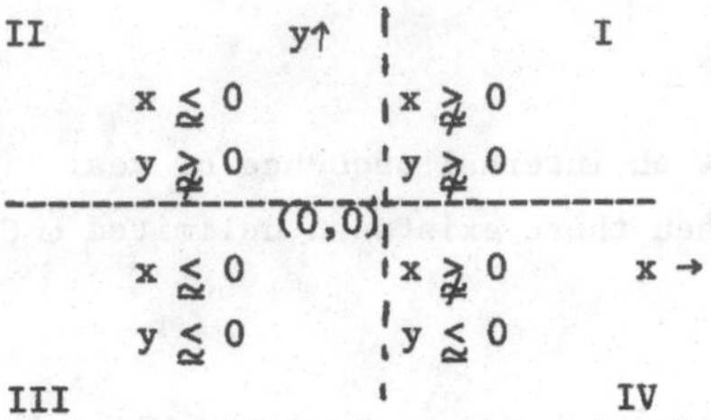

It is very well possible to walk along a continuous internal line from region III to region I, without touching region II or IV. Indeed, one could follow the line y = x. But it is impossible to follow a continuous internal line from region II into region IV, without passing through one of the regions I or III. In order to see this, let us consider the slightly more general situation:

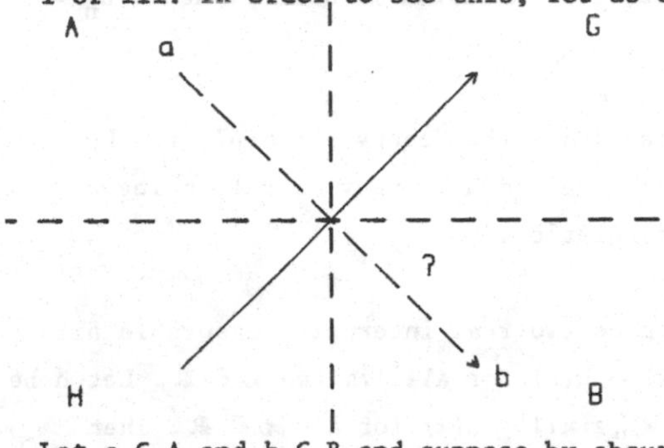

Going upside down, the horizontal dotted line suggest a frontier galaxy-halo, and going from the left to the right, the vertical dotted line suggest a frontier halo-galaxy. So G is a galaxy, H is a halo, A is the Cartesian product $p_x(H) \times p_y(G)$ and B is the Cartesian product $p_x(G) \times p_y(H)$.

Let $a \in A$ and $b \in B$ and suppose by absurdity that $f: [0,1] \to \mathbb{R}^2$ is a continuous function such that $f(0) = a$ and $f(1) = b$, and such that $\operatorname{im} f \cap H = \emptyset$ and $\operatorname{im} f \cap G = \emptyset$. Then $\operatorname{im} f \cap A$ is a prehalo for $\operatorname{im} f \cap A = \operatorname{im} f \cap (p_x(H) \times \mathbb{R})$; it follows from Theorem 4.9 that it is external, i.e. a halo. Likewise, $\operatorname{im} f \cap A$ is a pregalaxy, for $\operatorname{im} f \cap A = \operatorname{im} f \cap (\mathbb{R} \times p_y(G))$; as a consequence of Theorem 4.9, it is a galaxy. So we obtained a contradiction to the Fehrele principle. Hence every continuous way from A to B necessarily touches G or H.

5. <u>External order structure of $\mathbb{R}$.</u>

<u>Definition 4.13</u>: A set E is called <u>externally finite</u> if its cardinal is a standard natural number. A set E is called <u>externally countable</u> it there exists a one-to-one mapping onto some subset of $\underline{\mathbb{N}}$.

<u>Notation</u>: If A and B are two subsets of an ordered set such that $a < b$ for all $a \in A$ and $b \in B$, then we will write $A < B$.

We recall that the order relation on $\mathbb{R}$ has the property of being <u>dense</u>: between two real numbers x and z such that $x < z$ there exists some y such that $x < y < z$. An ordering on some set S is sometimes called to possess the <u>η_0-property</u> with respect to this set if the ordering is dense, and S possesses no minimum and no maximum. Likewise, an ordering on a set S is said to possess the <u>η_1-property</u> with respect to this set if for each pair of countable subsets A and B such that $A < B$ there exists $v \in S$ such that $A < v < B$, and if for each countable subset $V \subset S$ there exist a and b in S such that a < V < b. For higher ordinals α one may define η_α-properties in analogous way. These notions were introduced by Hausdorff, and reflect the richness of the ordered structure. As an example, the set of real numbers satisfies the η_0-property but not the η_1-property. But its external order structure is much richer: it satisfies a property which we may call the external η_1-property: for each pair of externally countable subsets A and B such that A < B there exists $v \in \mathbb{R}$ such that A < v < B, and for each externally countable subset V, there exists $a,b \in \mathbb{R}$ such that a < V < b. Indeed, the first property follows from the Fehrele principle, and the latter from the Cauchy principle. We will see in section 3 that $\mathbb{R}$ possesses what we may call the external η_α-property for each ordinal α.

6. The Fehrele principle has been very useful in the problem of matching local approximations of solutions of singularly perturbed differential equations. Its use has been explained by M. Diener in [25].

Classical results related to the Cauchy principle and the Fehrele principle.

Classical results related to nonstandard permanence phenomena in $\mathbb{R}$ concern the behaviour of real functions. This is not unexpected, in view of the possibility to carry out nonstandard analysis in the structure $\mathbb{R}^{\mathbb{R}}/U$ where U is some free ultrafilter. Classical permanence results originate from work of notably Du Bois-Reymond and Hadamard. An account of these results can be found in the book "Orders of infinity" of Hardy [46]. We will mention some of them. Let the partial order relation $\ll$ be defined by

$$f \ll g \leftrightarrow f = o(g) \text{ for } x \to \infty \leftrightarrow \lim_{x \to 0} f(x)/g(x) = 0$$

I. Classical results related to the Cauchy principle.

1) If $(f_n)_{n\in\mathbb{N}}$ is an increasing sequence of real functions, then there exists a real function g such that $f_n \ll g$ for all $n \in \mathbb{N}$ (Du Bois-Reymond's lemma).

2) If $(f_n)_{n\in\mathbb{N}}$ is an increasing sequence of real functions and h is a real function such that $f_n \ll g$ for all $n \in \mathbb{N}$ then there exists a real function g such that $g \ll h$ and $f_n \ll g$ for all $n \in \mathbb{N}$.

II. Classical results related to the Fehrele principle.

1) If $(f_n)_{n\in\mathbb{N}}$ is an increasing sequence of real functions and $(h_n)_{n\in\mathbb{N}}$ is a decreasing sequence of real functions such that $f_n \ll h_n$ for all $n \in \mathbb{N}$, then there exists a real function g such that $f_n \ll g \ll h_n$ for all $n \in \mathbb{N}$.

2) If $(f_n)_{n\in\mathbb{N}}$ is an increasing sequence of real functions such that $\lim_{x\to\infty} f_n(x) = 0$ for all $n \in \mathbb{N}$, then there exists a real function g such that $\lim_{x\to\infty} g(x) = 0$ and $f_n \ll g$ for all $n \in \mathbb{N}$.

We see that the order relation $\ll$ on the function set $\mathbb{R}^{\mathbb{R}}$ possesses the η_1-property. This property is evidently shared by other partial order relations, like the relation, say R, defined by $f\ R\ g \Leftrightarrow (\exists A)(\forall x \geq A)(f(x) < g(x))$.

Du Bois-Reymond has also proved a sequential lemma which can be seen as a classical counterpart of Robinson's lemma: if $(a_{n,k})_{n,k\in\mathbb{N}}$ is a double sequence such that $\lim_{n\to\infty} a_{n,k} = 0$ for all $k \in \mathbb{N}$, then there exists a sequence $(k_n)_{n\in\mathbb{N}}$ such that $k_n \to \infty$ and still $\lim_{n\to\infty} a_{n,k_n} = 0$. See an article of Laugwitz [33].

Finally, we mention Eckhaus' extension principle, based on Du Bois-Reymond's lemma, and used in connection with matching problems in classical perturbation theory [44].

3. <u>Generalized galaxies and halos</u>.

<u>Classification of external sets</u>.

<u>Remark</u>: We always assume that external quantifiers are relativized; for instance, if we write $\exists^{st} x\, A(x)$ it is tacitly assumed that there is some standard set X such that $\exists^{st} x\, A(x) \Leftrightarrow (\exists^{st} x)(x \in X \wedge A(x))$.

<u>Theorem 4.14</u>: Every external set E in IST can be reduced to one of the following three types:

1. Sets defined with one existential external quantifier: $z \in E \Leftrightarrow \exists^{st} x\, A(x,z)$, with A internal.

2. Sets defined with one universal external quantifier: $z \in E \Leftrightarrow \forall^{st} x\, A(x,z)$, with A internal.

3. Sets defined with one universal external quantifier, followed by one existential external quantifier: $z \in E \Leftrightarrow \forall^{st} x\, \exists^{st} y\, A(x,y,z)$, with A internal.

The theorem is an immediate consequence of Nelson's reduction algorithm [5]; in particular a formula of the form $\exists^{st} u\, \forall^{st} v\, B(u,v,z)$ is shown to be equivalent to a formula of the form $\forall^{st} x\, \exists^{st} y\, A(x,y,z)$.

<u>Definition 4.15</u>: Let X be a standard set and $(A_x)_{x\in X}$ be an internal family of internal sets. Then $\bigcup_{x\in X} A_x$ will be called a <u>generalized pregalaxy</u>. A generalized pregalaxy which is external will be called a <u>generalized galaxy</u>. A generalized galaxy which is not a galaxy will be called <u>strictly generalized</u>. In the same manner,

the set $\bigcap_{x \in X} A_x$ will be called a generalized prehalo. A generalized prehalo which is external will be called a generalized halo. A generalized halo which is no halo will be called strictly generalized.

So sets which fall in the first category of the above theorem are generalized galaxies, and sets of the second category are generalized halos. For sets of the third category the name "galo" is sometimes proposed.

It is to be expected that sets of one of the above types appear less frequently in "nature" then ordinary galaxies and halos. But there are natural examples, even in asymptotics (set of numbers possessing a shadow expansion with respect to some order scale, universal sum - see Chapter VI).
Furthermore, their study is motivated by the fact that there exist only so few types of them, and that fundamental results are rather easily proved.

In this section we show that no generalized halo is a generalized galaxy. In fact, a still more general statement about permanence can be proved (Theorem 4.16). We conclude this section of descriptive set theory with a further subdivision of the class of generalized galaxies.

General permanence theorems.

Theorem 4.16: (General permanence principle). Let X and Y be standard sets. Let $R \subset X \times Y$ be an internal relation such that $R(x,y)$ for all $x \in \underline{X}$ and $Y \in \underline{Y}$. Then there exist internal sets $I \subset X$ and $J \subset Y$ such that $R(x,y)$ for all $x \in I$ and $y \in J$.

Proof: We consider the internal relation $B(u,P)$ defined by

$$B(u,P) \Leftrightarrow u \in P \wedge P \subset R \wedge P \text{ is a Cartesian product.}$$

The result will be a consequence of the idealization principle, applied to the formula B. Indeed, let $z \subset X \times Y$ be a standard finite set. Let p be the projection of $X \times Y$ onto X and q be the projection of $X \times Y$ onto Y. Put $P_z = p(z) \times q(z)$. Then $P_z \subset \underline{X} \times \underline{Y} = \underline{X \times Y} \subset R$. Hence $B(u,P_z)$ is verified by all $u \in z$. By the idealization principle, there exists an internal set P such that $B(u,P)$ for all $u \in \underline{X \times Y}$. Then P is a rectangle $I \times J$, where I and J are internal sets such that $\underline{X} \subset I$, $\underline{Y} \subset J$ and $I \times J \subset R$.

With external induction, this result can be extended to relations of n variables, for all st n. The result is useful when permanence is to be applied in more than one dimension, and has indeed been used in this sense ([10] and application below). With the aid of the general permanence principle, we now prove that the classes of the generalized galaxies and the generalized halos are disjoint.

Theorem 4.17: (i) Let G be a generalized galaxy and H be a generalized halo such that $G \subset H$. Then there exists an internal set S such that $G \subset S \subset H$.

(ii) No generalized galaxy is a generalized halo.

Proof: (i) Let X and Y be two standard sets and $(A_x)_{x \in X}$ and $(B_y)_{y \in Y}$ be two internal families such that $G = \bigcup_{x \in \underline{X}} A_x$ and $H = \bigcap_{y \in \underline{Y}} B_y$. We define an internal relation $R \subset X \times Y$ by

$$R(x,y) \Leftrightarrow A_x \subset B_y.$$

Then $\underline{X \times Y} \subset R$. By the general permanence principle, there exist internal sets $I \supset \underline{X}$ and $J \supset \underline{Y}$ such that $I \times J \subset R$. So

$$G = \bigcup_{x \in \underline{X}} A_x \subset \bigcup_{x \in I} A_x \subset \bigcap_{y \in J} B_y \subset \bigcap_{y \in \underline{Y}} B_y = H$$

Putting, for instance, $S = \bigcup_{x \in I} A_x$, we obtain an internal set S such that $G \subset S \subset H$.

(ii) Let G, H and S be as above. By the Cauchy principle $G \subsetneq S \subsetneq H$. Hence $G \neq H$.

Comment: In logic it is customary to call Σ_1-sets those sets S which can be defined by a formula with only one existential quantifier, i.e. $x \in S \Leftrightarrow \exists y\, A(x,y)$, where A is a formula without quantifiers. Likewise sets which can be defined by a formula with only one universal quantifier are called Π_1-sets. Sets which are both Σ_1 and Π_1 are called Δ_1-sets.

In the same spirit generalized pregalaxies could be called "Σ_1^{st} sets" and generalized prehalos "Π_1^{st} sets". Then Theorem 4.17(ii) could be restated to "Sets which are Δ_1^{st} are internal".

Examples of generalized galaxies and halos. 1. Let $\omega \in \mathbb{R}$ be positive and unlimited. Let F be the set of all increasing real functions. Let $G_\omega \subset \mathbb{R}^+$ be the generalized pregalaxy, defined by

$$x \in G_\omega \Leftrightarrow 0 \leqq x \leqq f(\omega) \text{ for some } f \in \underline{F}$$

We will show that G_ω is a strictly generalized galaxy.

(i) G_ω is external: it is easy to see that G_ω is an additive convex semigroup. Now the only internal convex semigroups included in $\mathbb{R}^+$ are $\{0\}$ and $\mathbb{R}^+$. Clearly, $G_\omega \neq \{0\}$. On the other hand, it follows from an application of the idealization principle to the formula $B(f,y)$ defined by

$$B(f,y) \leftrightarrow f \in F \wedge f(\omega) < y$$

that there exists $y \in \mathbb{R}^+$ such that $y > G_\omega$. Hence $G_\omega \neq \mathbb{R}^+$. Consequently G_ω is external.

(ii) G_ω is no galaxy: if G_ω were a galaxy we could find standard functions $(f_n)_{n\in\underline{\mathbb{N}}}$ in F such that $G_\omega = \bigcup_{n\in\underline{\mathbb{N}}} [0,f_n(\omega)]$. By the standardization principle, the sequence $(f_n)_{n\in\underline{\mathbb{N}}}$ is the initial part of a standard sequence $(f_n)_{n\in\mathbb{N}}$. By the lemma of Du Bois-Reymond there exists a function $g \in F$ such that for all $n \in \mathbb{N}$, there exists $a_n \geqq 0$ such that $g(x) > f_n(x)$ for all $x \geqq a_n$. By transfer, the function g and the sequence $(a_n)_{n\in\mathbb{N}}$ may be assumed to be standard. This implies a contradiction, for $g(\omega) \in G_\omega$ and $g_\omega > f_n(\omega)$ for all st n. Hence G_ω is not a galaxy.

2. All sets which contain only standard elements and which are not standard finite are generalized galaxies. Those who are not externally countable are strictly generalized galaxies. This follows from next proposition, which implies that generalized galaxies containing only standard elements may be classified by cardinality.

Proposition 4.18: Let X and Y be two standard infinite sets such that $\mathrm{card}(X) < \mathrm{card}(Y)$. Then there does not exist an internal family $(A_x)_{x\in X}$ such that $\underline{Y} = \bigcup_{x\in\underline{X}} A_x$.

Proof: Suppose $(A_x)_{x\in X}$ is an internal family such that $Y = \bigcup_{x\in\underline{X}} A_x$. For all st x the internal sets A_x contain only standard elements. So they are externally finite, hence standard. Let $(A'_x)_{x\in X}$ be the standardized of $(A_x)_{x\in\underline{X}}$. By transfer, $Y = \bigcup_{x\in X} A'_x$, and all sets A'_x are finite. Now there exists a bijection from X onto $\bigcup_{x\in X} A'_x$, hence onto Y. This contradicts the assumption card X < card Y.

3. On every interval $[\omega,\infty)$, where ω is positive unlimited, there exists an internal function which separates the standard bounded functions from the standard functions going off to positive infinity. Indeed, let B be the set of all bounded real functions on $[0,\infty[$ and P be the set of all real functions f such that $\lim_{x\to\infty} f(x) = +\infty$. Then a standard element of B takes only limited values for unlimited arguments and a standard element of P takes only unlimited values for unlimited arguments.

For all $f \in \underline{P}$, let $a_f = \inf_{x \geq \omega} f(x)$. Then a_f is positive unlimited. So $H = \bigcap_{f\in\underline{P}} [0,a_f]$ is a generalized prehalo which strictly includes the galaxy $\mathbb{C}^+$. Let $a \in H-\mathbb{C}^+$ be arbitrary. Then the function $I: [\omega,\infty] \to \mathbb{R}$ defined by $I(x) = a$ will do.

We cannot ask for an internal function I which separates $\underline{B}$ and $\underline{P}$ for all positive unlimited arguments. Indeed, without restriction of generality we may assume I increasing. By the Cauchy principle, the domain of definition of I includes some interval $[c,\infty)$ where c is standard. Let the function $J: [c,\infty[\to \mathbb{R}$ equal ${}^{s}\{(x, {}^{o}(I(x))) \mid \text{st } x \in [c,\infty[\}$. Then $J \in \underline{B}$ or $J \in \underline{P}$, a contradiction.

4. Topological halos. It follows from comment 4 of section 2 that monads of points in topological spaces are generalized prehalos. Consider $\mathbb{R}$ with the usual topology. Monads of standard points are halos. But monads of nonstandard points can be strictly generalized halos. Let $\omega \in \mathbb{N}^+$ be unlimited. Let U be the monad of ω and U_ω be the convex component of U containing ω (the set U is not convex; to see this apply idealization to the formula $B(V,y) \Leftrightarrow V$ open neighbourhood of $\omega \wedge y \in V \wedge y > \omega+1$). Let S be the set of all sequences of strictly positive real numbers. Put $C_\omega = \bigcap_{s \in \underline{S}}]\omega - s_\omega, \omega + s_\omega[$. We will show that $U_\omega = C_\omega$.

Indeed, let $x \in U_\omega$, and $(s_n)_{n \in \mathbb{N}} \in \underline{S}$. Now $\bigcup_{n \in \mathbb{N}}]n-s_n, n+s_n[$ is a standard open neighbourhood of ω, so it contains x. Hence $U_\omega \subset C_\omega$. Conversely, let V be a standard open set containing ω. We may assume that every convex component of V has finite length. For all $n \in \mathbb{N}$, let V_n be the convex component of V containing n. (V_n may be empty). Define standard sequences $\varphi, \Psi: \mathbb{N} \to \mathbb{R}$ by

$$\varphi_n = \begin{cases} n - \inf V_n, & n \in V \\ 1 & \text{otherwise} \end{cases} \qquad \Psi_n = \begin{cases} \sup V_n - n & n \in V \\ 1 & \text{otherwise} \end{cases}$$

Then $]\omega - \varphi_\omega, \omega + \Psi_\omega[\subset V_\omega$. Hence $C_\omega \subset U_\omega$. Combining, we see that $U_\omega = C_\omega$. Now C_ω is a strictly generalized halo. This follows from example 1, for $\bigcap_{s \in \underline{S}} [0, s_\omega)$ is the image of $\mathbb{R}^+ - G_\omega$ under the inversion $x \to 1/x$.

5. $\mathbb{R}$ possesses the externally η_α-property for all ordinals α. This is an easy consequence of Theorem 4.17.

The following "multidimensional permanence principle" is an easy consequence of theorems 4.16 and 4.17.

Theorem 4.19: Let $n \in \mathbb{N}$ be standard, let $G_1, \dots, G_n$ be generalized pregalaxies and H be a generalized prehalo such that $G_1 \times \dots \times G_n \subset H$. Then there exist internal sets $I_1 \supset G_1, \dots, I_n \supset G_n$ such that $I_1 \times \dots \times I_n \subset H$.

Application: a polynomial distribution.

It is well-known that for small $\varepsilon > 0$, the function $x \to \frac{1}{\pi} \cdot \frac{\varepsilon}{\varepsilon^2 + x}$ behaves like a distribution. Indeed, suppose $\varepsilon \simeq 0$, and φ is an arbitrary standard continuous test function with compact carrier $[a,b]$, st $a < 0$, st $b > 0$. Then one gets, by an easy application of the lemma of dominated approximation

$$\int_{-\infty}^{\infty} \frac{1}{\pi}\cdot\frac{\varepsilon}{\varepsilon^2+x^2}\cdot\varphi(x)\,dx = \frac{1}{\pi}\int_{a/\varepsilon}^{b/\varepsilon} \frac{\varphi(\varepsilon y)}{1+y^2}dy \simeq \frac{1}{\pi}\int_{-\infty}^{\infty} \frac{\varphi(0)}{1+y^2}dy = \varphi(0)$$

Let us show that it is possible to define an explicit polynomial P(x) with the same property, i.e. for all standard testfunctions

$$\int_{-\infty}^{\infty} P(x)\varphi(x)\,dx \simeq \varphi(0)$$

Let $s > 0$ be appreciable, and $\omega \in \mathbb{N}$ be positive unlimited. Using the results from chapter III concerning the approximation of the exponential function by its Taylor-polynomials, one shows that the following approximation is valid for all limited x:

$$1 \simeq 1-e^{-\omega(\frac{s^2+x^2}{s})} = \sum_{n=1}^{\omega^2} (-1)^{n-1} \frac{\omega^n}{n!}\left(\frac{s^2+x^2}{s}\right)^n$$

Dividing both sides by $(s^2+x^2)/s$, one gets

$$\frac{s}{s^2+x^2} \simeq \sum_{n=1}^{\omega^2} (-1)^{n-1} \frac{\omega^n}{n!}\left(\frac{s^2+x^2}{s}\right)^{n-1}$$

By Theorem 4.19, there exist positive $\varepsilon \simeq 0$, negative unlimited X^- and positive unlimited X^+ such that for all $x \in [X^-,X^+]$

$$\frac{\varepsilon}{\varepsilon^2+x^2} \simeq \sum_{n=1}^{\omega^2}(-1)^{n-1} \frac{\omega^n}{n!}\left(\frac{\varepsilon^2+x^2}{\varepsilon}\right)^{n-1}$$

Put $P(x) = \sum_{n=1}^{\omega^2} (-1)^{n-1} \frac{\omega^n}{n!}\left(\frac{\varepsilon^2+x^2}{\varepsilon}\right)^{n-1}$. Then P(x) has the required property.

The existence of a minimal "duration" of permanence.

Consider the following problem. Let X be a standard infinite set, and let P be an internal property, which is verified for all standard elements of X. By the Cauchy principle, we know that $\{x \in X \mid P(x)\}$ is an internal set, which strictly contains the external set $\underline{X}$. Suppose one "quits" the set $\underline{X}$ without returning. To what extent is the property P still satisfied?

Let us precise the notion of "quitting" in the following manner. A natural way to leave $\underline{X}$ is to follow a standard one-to-one sequence $s = (s_n)_{n\in\mathbb{N}}$ of elements of X (then s_n is necessarily standard for st n and nonstandard for nonstandard n). Now by the Cauchy principle, there exists unlimited ω_s such that $P(s_n)$ for all $n \leqq \omega_s$. Does there exist a "uniform duration of permanence", i.e. some unlimited index ρ such that $P(s_n)$ for all standard sequences s and all $n \leqq \rho$? The answer is yes, on behalf of the following "universal index principle".

Proposition 4.20 (universal index principle). Let X be a standard set, and $f: X \to \mathbb{N}$ be an internal mapping such that f(x) is unlimited for all standard $x \in X$. Then there exists some unlimited $\rho \in \mathbb{N}$ such that $\rho < f(x)$ for all $x \in \underline{X}$.

Proof: By Theorem 4.17, the galaxy $\underline{\mathbb{N}}$ is strictly included in the generalized prehalo $\bigcap_{x \in \underline{X}} \{0,\dots,f(x)\}$. Hence there is some unlimited ρ preceding f(x) for all st x.

The proposition is useful when permanence is repeatedly applied. As an example of such use, this proposition helps us to define a "sum" to all shadow expansions with respect to some order scale (Theorems 6.12, 6.13 and 6.15 and comment).

For internal relations in two variables the problem could be stated as follows. Suppose X and Y are standard sets such that $\underline{X \times Y} \subset R \subset X \times Y$. Does there exist some unlimited index ρ such that $R(s_n,t_m)$ for all standard sequences $(s_n)_{n\in\mathbb{N}}$ and $(t_m)_{m\in\mathbb{N}}$? Next theorem gives a positive answer to this question. The theorem is easily generalized for relations with a standard number of variables.

Theorem 4.21: Let X and Y be standard sets and let R be an internal relation such that R(x,y) for all $(x,y) \in \underline{X \times Y}$. Then there exists an unlimited index ρ such that $R(s_n,t_m)$ for all standard sequences $(s_n)_{n\in\mathbb{N}}$ on X and $(t_m)_{m\in\mathbb{N}}$ on Y and for all $n,m \leqq \rho$.

Proof: Let S be the set of all sequences of elements of X and T be the set of all sequences of elements of Y. Define the internal relation $U \subset \mathbb{R} \times \mathbb{N} \times T \times \mathbb{N}$ by

$$U(s,n,t,m) \Leftrightarrow R(s_n,t_m)$$

Clearly, $U \supset \underline{S} \times \underline{\mathbb{N}} \times \underline{T} \times \underline{\mathbb{N}}$. By the general permanence principle, applied to a relation of four variables, there are internal sets $I \supset \underline{S}$ and $J \supset \underline{T}$ and unlimited numbers ρ_1 and ρ_2 such that $U \supset I \times \{0,\dots,\rho_1\} \times J \times \{0,\dots,\rho_2\}$. Let ρ be any number less than ρ_1 and ρ_2. Then ρ has the required property.

Increasing and disjoint representation.

We conclude this section with a further subdivision of the class of external sets. We will distinguish generalized galaxies with "increasing representation" and with "disjoint representation".

Definition 4.22: Let G be a generalized galaxy. We say that G possesses an increasing representation if $G = \bigcup_{x\in \underline{X}} A_x$, where X is a totally ordered standard set and $(A_x)_{x\in X}$ is an internal family such that $x_1 < x_2$ implies $A_{x_1} \subset A_{x_2}$, for all $x_1, x_2 \in \underline{X}$. We say that G possesses a disjoint representation if $G = \bigcup_{y\in \underline{Y}} B_y$, where Y is a standard set and $(B_y)_{y\in Y}$ is an internal family such that $B_{y_1} \cap B_{y_2} = \emptyset$ for all $y_1, y_2 \in \underline{Y}$.

Theorem 4.23: (i) Every galaxy has an increasing representation and a disjoint representation.

(ii) A strictly generalized galaxy cannot have both an increasing representation and a disjoint representation.

An example of a strictly generalized galaxy with increasing representation is given by the set $G_\omega = \bigcup_{f\in \underline{\mathbb{R}}^{\mathbb{R}}} [0, f(\omega)]$ of example 1 above. The sets containing the standard elements of uncountable standard sets are examples of strictly generalized galaxies with disjoint representation.

Theorem 4.22(i) is easily proved. By definition, a galaxy has an increasing representation. By proposition 4.8(i), a galaxy G may also be written in the form $G = f^{-1}(\underline{\mathbb{N}})$ where f is some internal function. Hence G possesses a disjoint representation. Part (ii) of the theorem will be proved with the help of next lemma.

Lemma 4.24: Let F be a pregalaxy included in a strictly generalized galaxy G which possesses an increasing representation. Then there exists an internal set I such that $F \subset I \subset G$.

Proof: Let $\bigcup_{n\in \underline{\mathbb{N}}} A_n$ be an increasing representation of F and $\bigcup_{x\in \underline{X}} B_x$ be an increasing representation of G. For every standard $n \in \mathbb{N}$, choose $x_n \in \underline{X}$ such that $A_n \subset B_{x_n}$. Now G is not a galaxy, so there exists $\xi \in \underline{X}$ such that $B_\xi \not\subset \bigcup_{n\in \underline{\mathbb{N}}} B_{x_n}$. Because $\{B_x \mid x \in \underline{X}\}$ is totally ordered by inclusion, one has $\bigcup_{n\in \underline{\mathbb{N}}} B_{x_n} \subset B_\xi$. Putting $I = B_\xi$, one finds

$$F = \bigcup_{n\in \underline{\mathbb{N}}} A_n \subset \bigcup_{n\in \underline{\mathbb{N}}} B_{x_n} \subset I \subset G$$

Proof of Theorem 4.23(ii): Let G be a strictly generalized galaxy with an increasing representation. Suppose G has also a disjoint representation $\bigcup_{x\in \underline{X}} A_x$. Without loss of generality we may assume that $A_x \neq \emptyset$ for all $x \in \underline{X}$. Let $(x_n)_{n\in \underline{\mathbb{N}}}$ be a one-to-one sequence of elements of $\underline{X}$ and let F be the galaxy $\bigcup_{n\in \underline{\mathbb{N}}} A_{x_n}$. By lemma 4.24, there is some internal subset I of G such that $F \subset I$. One even has $F \subsetneq I$, for F is external. Let H be the halo $I \setminus F$. Let $(B_x)_{x\in X}$ be the internal family defined by $B_x = A_x \cap I$. Then $H = \bigcup_{x\in \underline{X} - \{x_n \mid n\in \underline{\mathbb{N}}\}} B_x$, which contradicts theorem 4.17. So G cannot have a disjoint representation.

Finally we state an analogon to Robinson's sequential lemma for strictly generalized galaxies. The result follows easily from the above lemma.

Proposition 4.25: Let G be a strictly generalized galaxy which has an increasing representation. Let $(a_n)_{n \in \mathbb{N}}$ be an internal sequence such that $a_n \in G$ for all $n \in \underline{\mathbb{N}}$. Then $a_n \in G$ up to some unlimited $\omega \in \mathbb{N}$.

4. Convex external sets.

Convex external subsets of $\mathbb{R}$ may be thought of as genuine "asymptotic sets". Acting as domains of asymptotic behaviour and reasoning, as orders of magnitude and as models of continuous transitions, they are basic to our approach to asymptotics. Having no sharp boundaries, they provide great flexibility, but still order reigns: they can take only a few, definite forms, and possess clearly defined boundary sets, which are translations of convex additive groups.

We first study convex galaxies and halos of $\mathbb{R}$. Then we present a classification of external convex subsets of $\mathbb{R}$ in four definite types, this classification still holds true in arbitrary convex ordered groups. The classification theorem will be the starting point to define "solutions" to the "external equations" of section 5.

Convex galaxies and halos in $\mathbb{R}$.

As will be shown later, convex galaxies and halos in $\mathbb{R}$ may be written in terms of convex subgroups of $\mathbb{R}$. So we first turn our attention to convex galaxies and halos of this form, starting with galaxies. The following elementary proposition is very helpful in this context.

Proposition 4.26: (i) A subset G of $\mathbb{R}$ is a convex galaxy which is symmetric with respect to 0 if and only if there exists an internal strictly increasing sequence of strictly positive real numbers $(a_n)_{n \in \mathbb{N}}$ such that $G = \bigcup_{n \in \underline{\mathbb{N}}} [-a_n, a_n]$.

(ii) A subset G of $\mathbb{R}$ is a convex galaxy which is symmetric with respect to 0 if and only if there exists a real internal strictly increasing odd C^{∞} function f such that $f(\pounds) = G$.

Examples. 1. The α-galaxies (see example 4.2.2) are convex additive subgroups of $\mathbb{R}$. They are the image of the principal galaxy under the linear function $f(x) = \alpha x$. They are sometimes called linear galaxies. Convex additive groups which are not α-galaxies are sometimes called nonlinear galaxies. The set $m_\varepsilon = \{x \mid |x| = e^{-1/(b\varepsilon)}, b \geq 0$ limited$\}$ is such a galaxy. Indeed, if $\alpha \in m_\varepsilon$, then $\sqrt{|\alpha|} \in m_\varepsilon$, and $\sqrt{|\alpha|}$ is certainly not of order α. The set m_ε often occurs in connection with singularly perturbed differential equations (for instance, it is the size of the set of values of the parameter a, for which the "canard" phenomenon appears in the Van der Pol equation $\varepsilon\ddot{x}+(x^2-1)\dot{x}+x = 1-a$ [11] ; it also acts as "thickness of jump", see below) and has been called the <u>ε-microgalaxy</u>.

2. (this example has been taken from [21]). Solutions of singularly perturbed differential equations may produce <u>jumps</u>, i.e. during an infinitesimal time-lapsus, the solution x(t), though continuous moves from some niveau x_- to some niveau x^+, with $x_- \not\simeq x^+$ (x_- and x^+ are supposed to be standard).

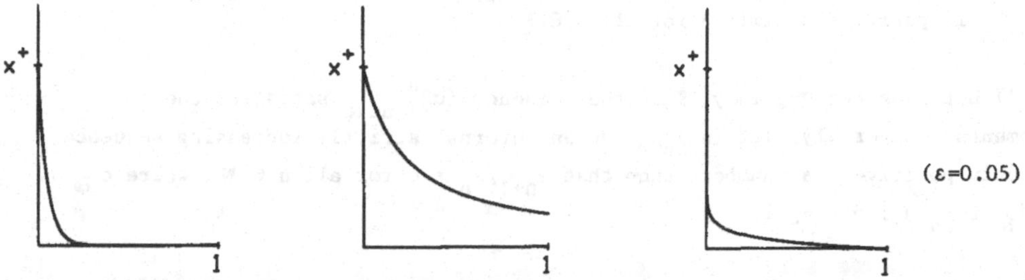

$\varepsilon\ddot{x} = -\dot{x}$ $\quad$ $\varepsilon\ddot{x} = 2x^3$ $\quad$ $\varepsilon\ddot{x} = (\dot{x})^2$

$x(t) = e^{-t/\varepsilon}$ $\quad$ $x(t) = \sqrt{\varepsilon}/(t+\sqrt{\varepsilon})$ $\quad$ $x(t) = -\varepsilon\log(t+e^{-1/\varepsilon})$

The set $\{t \mid x_- \lneqq x(t) \lneqq x^+\}$, if $x_- < x^+$, or the set $\{t \mid x^+ \lneqq x(t) \lneqq x_-\}$, in the case $x^+ < x_-$, have been called the <u>thickness</u> of the jump, say T. If one knows this thickness, we can adjust the time, i.e. determine a C^∞ monotonous application $h\colon \mathbb{R} \to \mathbb{R}$ such that $h(\mathbb{C}) = T$ (or, say, $h(\mathbb{A}^+) = T$), in order to see what happens during the jump ("stretching"). Here are some examples ($\varepsilon \simeq 0$, $\varepsilon > 0$).

1) The function $x(t) = e^{-t/\varepsilon}$, $t \in [0,1]$ is solution of the equation $\varepsilon\ddot{x} = -\dot{x}$. Near 0 it jumps from $x_- = 1$ to $x^+ = 0$, and $T = \varepsilon\mathbb{A}^+$.

2) The function $x(t) = \sqrt{\varepsilon}/(t+\sqrt{\varepsilon})$, $t \in [0,1]$ is solution of the equation $\varepsilon\ddot{x} = 2x^3$. Near 0 it jumps from $x_- = 1$ to $x^+ = 0$, and $T = \sqrt{\varepsilon}\mathbb{A}^+$.

3) The function $x(t) = -\varepsilon\log(t+e^{-1/\varepsilon})$, $t \in [0,1]$ is solution of the equation $\varepsilon\ddot{x} = (\dot{x})^2$. Near 0 it jumps from $x^+ = 1$ to $x_- = 0$. Here $T = m_\varepsilon^+ - \{t \mid t = e^{-(1+\alpha)/\varepsilon}, \alpha \gtrsim 0\}$.

It is of some importance whether the time-adjustment h can be a linear mapping, (as in the first and second case), or is necessarily nonlinear (as in the latter case). Below we study how to recognize nonlinear galaxies from the linear ones.

Let G be a galaxy which is symmetric with respect to 0. The following propositions give criteria to decide whether G is a linear galaxy or a nonlinear galaxy.

Proposition 4.27: Let $G \subset \mathbb{R}$ be a convex galaxy which is symmetric with respect to 0. Then

1) G is an α-galaxy if and only if there exists an internal strictly increasing sequence of strictly positive real numbers $(a_n)_{n\in\mathbb{N}}$ with $G = \bigcup_{n\in\underline{\mathbb{N}}} [-a_n, a_n]$, such that $a_0 = \alpha$, and $a_{n+1}/a_n = c$ for all $n \in \underline{\mathbb{N}}$, where $c \gneqq 1$ is limited .

2) G is nonlinear if and only if there exists an internal strictly increasing sequence of strictly positive real numbers $(a_n)_{n\in\mathbb{N}}$ with $G = \bigcup_{n\in\underline{\mathbb{N}}} [-a_n, a_n]$ such that a_{n+1}/a_n is positive unlimited for all $n \in \underline{\mathbb{N}}$.

Proof: 1) Let G be the α-galaxy. Then the sequence $(\alpha 2^n)_{n\in\mathbb{N}}$ satisfies the requirements. Conversely, let $(a_n)_{n\in\mathbb{N}}$ be an internal strictly increasing sequence of strictly positive real numbers such that $a_{n+1}/a_n = c$ for all $n \in \underline{\mathbb{N}}$, where $c \gneqq 1$. Then $\bigcup_{n\in\underline{\mathbb{N}}} [-a_n, a_n] = a_0\text{-gal}$.

2) Let G be nonlinear and $(b_k)_{k\in\mathbb{N}}$ be a strictly increasing sequence of strictly positive real numbers such that $G = \bigcup_{k\in\underline{\mathbb{N}}} [-b_k, b_k]$. We define a subsequence $(a_n)_{n\in\underline{\mathbb{N}}}$ of $(b_k)_{k\in\underline{\mathbb{N}}}$ such that $a_n/a_{n+1} \simeq 0$ by external induction. Put $a_0 = b_0$. Suppose a_n is defined to be some b_k. G is a convex group, so it contains a_n-gal, and being nonlinear, it contains a_n-gal strictly. So there is some $b_k \in G\setminus a_n\text{-gal}$. Put a_{n+1} the smallest of these b_k. Then $a_n/a_{n+1} \simeq 0$. By the principle of extension there exists an internal prolongation $(a_n)_{n\in\mathbb{N}}$ of $(a_n)_{n\in\underline{\mathbb{N}}}$, which we may assume strictly increasing. This sequence has all the required properties.

Conversely, let $(a_n)_{n\in\mathbb{N}}$ be an internal increasing sequence of strictly positive real numbers such that a_{n+1}/a_n is positive unlimited for all $n \in \underline{\mathbb{N}}$. Put $G = \bigcup_{n\in\underline{\mathbb{N}}} [-a_n, a_n]$. Then G is nonlinear: if $\alpha \leqq a_n$ for some $n \in \underline{\mathbb{N}}$, then $\alpha/a_{n+1} \simeq 0$. Hence $\alpha\text{-gal} \subset G$ implies $\alpha\text{-gal} \subsetneqq G$.

Proposition 4.28: Let $G \subset R$ be a convex galaxy which is symmetric with respect to 0. Then

1) G is an α-galaxy if and only if there exists a real internal strictly increasing odd C^∞-function f such that $f(\mathbb{G}) = G$ and $f'(x)/f(x) = c$ for all $x \geqq 1$, where c is a positive appreciable number.

2) G is nonlinear if and only if there exists a real internal strictly increasing odd C^∞-function f such that $f(\mathbb{G}) = G$ and $f'(x)/f(x)$ is unlimited for all limited

$x \geqq 1$.

Proof: 1) The equivalence follows in an obvious way from proposition 4.27.1).

2) Let G be nonlinear. Let $(a_n)_{n\in\mathbb{N}}$ be an internal strictly increasing sequence of strictly positive numbers such that $G = \bigcup_{n\in\mathbb{N}} [-a_n, a_n]$ and $\frac{a_{n+1}}{a_n}$ is unlimited for all $n \in \mathbb{N}$. We define the functions f_n on $[n,n+1]$ by

$$f_n(x) = \begin{cases} a_n\left(\frac{a_{n+1}}{a_n}\right)^{x-n} & n \geqq 1 \\ a_1 t & n = 0 \end{cases}$$

Then f_n is internal, strictly increasing and C^∞ on $[n,n+1]$ for all $n \in \mathbb{N}$. Furthermore $f'_n(x)/f(x) = \log(a_{n+1}/a_n)$ so $f'_n(x)/f(x)$ is unlimited for all $x \in [n,n+1]$, where $n \geqq 1$, while $\bigcup_{n\in\mathbb{N}} f_n$ is continuous. By smoothening in a suitable way we can obtain a function f which conserves all these properties and is in addition C^∞ on $[0,\infty[$; we may suppose that the odd extension of f to $]-\infty,0]$ is still C^∞.

The converse is simple.

Proposition 4.29: Let $G \subset \mathbb{R}$ be a convex galaxy which is symmetric with respect to 0. Then

(i) G is nonlinear if and only if for all $a \in G$ there is some $b \in G$ such that b/a is positive unlimited.

(ii) G is nonlinear if and only if there exists some positive unlimited $\omega \in \mathbb{R}$ such that $\omega G = G$.

Proof: We only prove the second half of (ii).

By proposition 4.27 there exists a strictly increasing sequence of positive real numbers $(a_n)_{n\in\mathbb{N}}$ such that $G = \bigcup_{n\in\mathbb{N}} [-a_n, a_n]$ and a_{n+1}/a_n is unlimited for all $n \in \mathbb{N}$. By the Fehrele principle, there exists some unlimited $\omega \in \mathbb{R}$ such that $\omega \leq \frac{a_{n+1}}{a_n}$ for all st n. Now let $x \in G$ be arbitrary. Then $|x| \leq a_m$ for some st m. Then $|\omega x| \leqq \frac{a_{m+1}}{a_m} \cdot a_m = a_{m+1} \in G$. Hence $\omega x \in G$.

The converse is trivial.

We will now consider convex halos which are symmetric with respect to 0. We use the results obtained thus far about convex galaxies to deduce analogous results about convex halos. So we have the following analogon to proposition 4.26.

Proposition 4.30. (1) A subset H of $\mathbb{R}$ is a convex halo which is symmetric with respect to 0 if and only if there exists an internal strictly decreasing sequence of strictly positive real numbers $(a_n)_{n\in\mathbb{N}}$ such that $H = \bigcap_{n\in\underline{\mathbb{N}}} [-a_n, a_n]$.

(2) A subset H of $\mathbb{R}$ is a convex halo which is symmetric with respect to 0 if and only if there exists a real internal strictly increasing odd C^∞-function f such that $f(\text{hal}(0)) = H$.

The α-halos (example 4.2.3) are examples of convex halos which are symmetric with respect to 0. They are additive subgroups of $\mathbb{R}$. Convex additive subgroups which are halos, but not α-halos are called nonlinear halos or microhalos. An example of such a microhalo is the set $\varepsilon\text{-M} \equiv \bigcap_{\text{st } n\in\mathbb{N}} [-\varepsilon^n, \varepsilon^n]$ where $\varepsilon \simeq 0$, $\varepsilon > 0$. The microhalos play a fundamental part in nonstandard asymptotics. As will be shown in Chapter VI, a formal shadow expansion determines a number r, which is unique up to a microhalo.

The following proposition gives criteria to determine whether a convex halo which is symmetric with respect to 0 is a nonlinear halo. The proofs are very similar to the proofs of 4.27-4.29.

Proposition 4.31: Let $H \subset \mathbb{R}$ be a convex halo which is symmetric with respect to 0. Then

(i) H is nonlinear if and only if there exists an internal strictly decreasing sequence of positive real numbers $(a_n)_{n\in\mathbb{N}}$ such that $H = \bigcap_{n\in\underline{\mathbb{N}}} [-a_n, a_n]$ and $a_{n+1}/a_n \simeq 0$ for all $n \in \underline{\mathbb{N}}$.

(ii) H is nonlinear if and only if there exists a real internal strictly increasing odd C^∞-function f such that $f(\text{hal}(0)) = H$ and $f'(x)/f(x)$ is positive unlimited for all $x \in \mathbb{A}^+$.

(iii) H is nonlinear if and only if for all $a \notin H$ there exists some $b \notin H$ such that $b/a \simeq 0$.

(iv) H is nonlinear if and only if there exists some positive unlimited $\omega \in \mathbb{R}$ such that $\omega H = H$.

The above propositions reflect the fact that firstly, it is often easier to get information about galaxies "from within", and secondly, it is often easier to get information about halos "from the outside".

Classification of convex subsets of $\mathbb{R}$

Definition 4.32: Let S be a totally ordered set. A subset C such that $c \in C$ implies that $x \in C$ for all $x \leqq c$ is called a lower halfline. A lower halfline of the form $\{x \in S \mid x \leqq a\}$ is written $]-\infty,a]$. A subset D such that $d \in D$ implies that $y \in D$ for all $y \geqq d$ is called an upper halfline. An ordered pair (C,D) composed of a lower halfline C and an upper halfline D such that $C \cup D = S$ and $C \cap D = \emptyset$ is called a cut.

Theorem 4.33: Let S be a totally ordered standard set. An external halfline of S is either a generalized galaxy or a generalized halo.

Theorem 4.34: Let S be a totally ordered standard group. Let (G,H) be a cut of S into a generalized galaxy G and a generalized halo H. Then there exists a unique convex subgroup K of S and an element a of S such that

either $G =]-\infty,a] \cup a+K$ and K is a generalized galaxy
or $G =]-\infty,a] \setminus a+K$ and K is a generalized halo.

Corollary 4.35: Let C be a convex subset of $\mathbb{R}$. Then there exist two unique additive convex subgroups K and L of $\mathbb{R}$ and two real numbers a and b such that C takes exactly one of the following forms:

1) $C = [a,b] \cup (a+L^-) \cup (b+K^+)$
2) $C = [a,b] \cup (a+L^-) \setminus (b+K^-)$
3) $C = ([a,b] \setminus (a+L^+)) \cup (b+K^+)$
4) $C = ([a,b] \setminus (a+L^+)) \setminus (b+K^-)$

Comments: 1) Corollary 4.35 states a normal form for convex subsets of $\mathbb{R}$ (of course this notion may be extended to convex subsets of arbitrary ordered groups). The corollary may be seen as a generalization of the classical completeness theorem of $\mathbb{R}$, which implies that every internal convex subset of $\mathbb{R}$ takes exactly one of the forms [a,b], [a,b[,]a,b] or]a,b[, where a and b are real numbers, or positive or negative infinity. It is clear that these forms can be rewritten to fit the representations of Corollary 4.35, where K and L can, of course, only be equal to $\{0\}$ or $\mathbb{R}$. For instance, $]-\infty,1[= [0,1] \cup a+\mathbb{R}^- \setminus a+\{0\}$.
2) Here are some examples of external lower halflines of $\mathbb{R}$ written in the representation of Theorem 4.34.

1. $\{x \mid x \leqq s \text{ for some st } s \in \mathbb{R}\} =]-\infty,0] \cup 0+ \pounds$

2. $\{x \mid x \lessapprox 1\} =]-\infty,1] \setminus 1+\text{hal}(0)$

3. Let ω be some unlimited number. Then

$$\{x \mid x \in f(\omega) \text{ for some standard real function}\} =]-\infty,0] \cup 0+G_\omega$$

(The set G_ω is defined in example 1 of Section 3).

3) Theorem 4.34 holds true for both commutative and noncummutative ordered groups. However, for reasons of readability, we will confine us in the future to cummutative groups.

4) Though only formulated for the case that the lower halfline is a generalized galaxy, Theorem 4.34 concerns all possible external cuts. It contains the case that the lower halfline is a generalized halo by symmetry. But now we have mentioned all cases, by Theorem 4.33. This theorem states that halflines are "simple" in the hierarchy of external sets: they may be defined with only one external quantifier, and cannot be "galos".

5) As easily verified on the above examples, the element a of S mentioned in Theorem 4.34 is not unique. The element a can be replaced by any $b \in S$, as long as $b-a \in K$. So the cut is stable under translation of elements of K: the group K acts as the thickness of the border of the cut. Let us state the following definitions.

Definition 4.36: Let (G,H) be a cut of a totally ordered commutative standard group S.

(i) The set $\Delta = \{d \in S \mid d = h-g \text{ for some } h \in H \text{ and } g \in G\}$ is called the set of differences for the cut (G,H).

(ii) The set $K = \{z \in S \mid |z| < d \text{ for all } d \in \Delta\}$ is called the thickness of the border of the cut (G,H).

6) Results concerning the thickness of the border of cuts are also found in classical literature on ordered groups. See [57] for a survey.

Proofs:

I. Corollary 4.35 is a direct consequence of Theorem 4.34 and the classical completeness theorem of $\mathbb{R}$.

II. Towards a proof of Theorem 4.33. Firstly, we mention that an internal set I contained in a generalized galaxy $\bigcup_{x \in X} A_x$, where X is standard and $(A_x)_{x \in X}$ is an internal family, is in fact already contained in a finite union $\bigcup_{x \in z} A_x$, where z is some standard finite subset of X. This compactness property is an immediate consequence of the idealization principle.

Secondly, we will use the fact that a convex component of a generalized galaxy in some ordered set is itself a generalized galaxy.

<u>Notation</u>: Let S be a totally ordered set, let $T \subset S$ and $a \in T$. The convex component of a within T will be written T^a.

<u>Proposition 4.37</u>: Let S be a standard totally ordered set. Let $G \subset S$ be a generalized pregalaxy and $a \in G$. Then G^a is a generalized pregalaxy.

<u>Proof</u>: Let X be a standard set and $(A_x)_{x \in X}$ be an internal family of internal sets, such that $G = \bigcup_{x \in \underline{X}} A_x$. Let Z be the standard set of all finite subsets of X. For all $z \in Z$ we define the internal set B_z by

$$B_z = \bigcup_{x \in z} A_x$$

We will show that $G^a = \bigcup_{z \in \underline{Z}} B_z{}^a$. Indeed

$$\begin{aligned} u \in G^a &\Leftrightarrow [a,u] \subset G \\ &\Leftrightarrow [a,u] \subset \bigcup_{x \in \underline{X}} A_x \\ &\Leftrightarrow (\exists^{st\ fin} z \subset X)([a,u] \subset B_z) \\ &\Leftrightarrow (\exists z \in \underline{Z})(u \in B_z{}^a) \\ &\Leftrightarrow u \in \bigcup_{z \in \underline{Z}} B_z{}^a \end{aligned}$$

<u>Proof of Theorem 4.33</u>: Without loss of generality, let C be a lower halfline of S and let $a \in C$. By Nelson's reduction algorithm, it is possible to write C in the form

$$C = \bigcap_{x \in \underline{X}} \bigcup_{y \in \underline{Y}} I_{xy}$$

where X and Y are standard and $(I_{xy})_{x \in X, y \in Y}$ is an internal family of internal subsets of S. Let us define generalized pregalaxies G_x by

$$G_x = \bigcup_{y \in \underline{Y}} I_{xy}$$

By Proposition 4.37, every set $G_x{}^a$ is also a generalized pregalaxy. It follows from the convexity of C that

$$C = \bigcap_{x \in \underline{X}} G_x{}^a$$

Now one has two possibilities:

1) $(\exists x \in \underline{X})(G_x^a \subset C)$. Then $C = G_x^a$ and C, being external, is a generalized galaxy.

2) $(\forall x \in \underline{X})(\exists u \in G_x^a \setminus C)$. For all $x \in \underline{X}$, choose $u_x \in G_x^a \setminus C$. Extend the external function $(u_x)_{x\in\underline{X}}$ to an internal function $(u_x)_{x\in X}$. Then $C = \bigcap_{x\in\underline{X}}]-\infty,u_x]$. Hence C is a generalized halo.

III. Towards a proof of Theorem 4.34. We first prove two lemma's.

Lemma 1. Let S be a totally ordered commutative standard group, and K be the thickness of the border associated to a cut (G,H) of S into a generalized galaxy G and a generalized halo H. Then (i) $G = \bigcup_{g\in G} g+K$, (ii) $H = \bigcup_{h\in H} h+K$, (iii) K is a convex subgroup of S.

Proof: (i) and (ii) are evident. (iii) The set K is not empty for $0 \in K$. Clearly, K is convex. Finally, let $z,w \in K$ and $g \in G$. Then $g+|z| \in G$. So $g+(|z|+|w|) = (g+|z|)+|w| \in G$. Hence $|z|+|w| \in K$. This implies that $z-w \in K$, for $|z-w| \leq |z|+|w|$. Hence K is a convex subgroup of S.

Let the generalized galaxy G be a lower halfline of S. Suppose $G = \bigcup_{x\in\underline{X}} A_x$, where X is a standard set and $(A_x)_{x\in X}$ is an internal family of subsets of S. Then there exists an internal function $f: X \to S$ such that G may be written as a union of halflines $\bigcup_{x\in\underline{X}}]-\infty,f(x)]$. Indeed, by the Cauchy principle the internal sets A_x are strictly included in G for all $x \in \underline{X}$. So choose for all $x \in \underline{X}$ an element $f(x) \in G$ which is larger than all elements of A_x. Then $G = \bigcup_{x\in\underline{X}}]-\infty,f(x)]$. By the extension principle the external function $f: \underline{X} \to S$ may be extended to some internal function $f: X \to S$.

Lemma 2. Let S be a totally ordered commutative standard group and let K be the thickness of the border associated to a cut (G,H) of S into a generalized galaxy G and a generalized halo H. Assume $G = \bigcup_{x\in\underline{X}}]-\infty,f(x)]$, where $f: X \to S$ is some internal function. Then there are two possibilities:

either there exists an internal function $\varphi: X \to S$ such that $K = \bigcup_{x\in\underline{X}}]-\varphi(x),\varphi(x)]$,

or there exists an internal function $\Psi: X \to S$ such that

$K = \bigcap_{x\in\underline{X}} [-\Psi(x),\Psi(x)]$ and $\Psi(x) \in \Delta$ for all $x \in \underline{X}$.

Proof: First suppose that G is of the form $[-\infty,a] \cup a+K$ for some $a \in G$. Define the internal mapping $\varphi: X \to S$ by

$$\varphi(x) = \begin{cases} f(x)-a & a \leq f(x) \\ 0 & \text{otherwise} \end{cases}$$

Then $K = \bigcup_{x\in\underline{X}} [-\varphi(x),\varphi(x)]$.

Secondly, suppose that there is no $a \in A$ such that $G =]-\infty,a] \cup a+K$. Then for all $u \in G$ there is some $v \in G$ such that $v > u$ and $v-u \notin K$, i.e. $v-u \in \Delta$. In particular, for all $x \in \underline{X}$ we may choose $y(x) \in \underline{X}$ such that $f(y(x))-f(x) \in \Delta$. Let $y: X \to X$ be the standardized of $y: \underline{X} \to \underline{X}$. Let $\Psi: X \to S$ be the internal function defined by

$$\Psi(x) = f(y(x))-f(x)$$

Then $\Psi(x) \in \Delta$ for all $x \in \underline{X}$. We will show that $K = \bigcap_{x\in\underline{X}} [-\Psi(x),\Psi(x)]$. On one hand $K \subset \bigcap_{x\in\underline{X}} [-\Psi(x),\Psi(x)]$ for $\varphi(x) \in \Delta$ for all $x \in \underline{X}$. On the other hand, assume $|z| \leqq \Psi(x)$ for all $x \in \underline{X}$. Let $g \in G$ be arbitrary, and let $\xi \in \underline{X}$ be such that $g < f(\xi)$. Then

$$g+|z| \leqq f(\xi)+\Psi(\xi) = f(\xi)+(f(y(\xi))-f(\xi)) = f(y(\xi)) \in G$$

Hence $z \in K$. So $\bigcap_{x\in\underline{X}} [-\Psi(x),\Psi(x)] \subset K$. We conclude that $K = \bigcap_{x\in\underline{X}} [-\Psi(x),\Psi(x)]$.

<u>Proof of Theorem 4.34</u>: Let Δ, K, X and $f: X \to S$ be as above.

1) <u>Existence</u>. Suppose there exists $a \in G$ such that $G =]-\infty,a] \cup a+K$. Then K is a convex group by lemma 1, and a generalized pregalaxy by lemma 2. Furthermore K is external for G is external. Hence K is a generalized galaxy. Now suppose that G is not of the form $]-\infty,a] \cup a+K$. By lemma 2 there exists an internal function $\Psi: X \to S$ such that $K = \bigcap_{x\in\underline{X}} [-\Psi(x),\Psi(x)]$ and $\Psi(x) \in \Delta$ for all $x \in \underline{X}$. So K is a generalized prehalo. Furthermore, for each $x \in \underline{X}$ we may choose $h(x) \in H$ such that $h(x)-\Psi(x) \in G$. Let $h: X \to S$ be an internal prolongation of $h: \underline{X} \to S$. Because the generalized galaxy G must be strictly included in the generalized prehalo $\bigcap_{x\in\underline{X}}]-\infty,h(x)]$, there exists some $a \in H$ such that $a < h(x)$ for all $x \in \underline{X}$. We will show that $G =]-\infty,a] \setminus a+K$.

On one hand $G \subset]-\infty,a] \setminus a+K$, for $a+K \subset H$ by lemma 1(ii). On the other hand, suppose $u < a+z$ for all $z \in K$. Then $u \leqq a-\Psi(\xi)$ for some $\xi \in \underline{X}$. Now $a-\Psi(\xi)<h(\xi)-\Psi(\xi) \in G$. So $u \in G$, implying $]-\infty,a]\setminus a+K \subset G$. We conclude that $G = -]-\infty,a] \setminus a+K$. This implies that K is external, and thus a generalized halo.

2) <u>Unicity</u>. Firstly, suppose that K' is a convex subgroup of S such that $K' \subsetneqq K$. Then it easily follows from lemma 1(i) and (ii) that K' would not do.

Secondly, let K'' be a convex subgroup of S such that $K \subsetneqq K''$. Let d be a positive element of $K''-K$. Then $d \in \Delta$. Let $g \in G$ and $h \in H$ be such that $g+d = h$. Suppose that G were of the form $]-\infty,a] \cup a+K''$. Then $a < g$. So $g-a \in K''$, and also $g-a+d \in K''$, for K'' is a group. But $a+(g-a+d) = h \in H$. This is a contradiction, so G is not of the form $]-\infty,a] \cup a+K''$. In the same manner one proves that G cannot be of the form $]-\infty,b] \setminus b+K''$ for some $b \in H$. Hence K'' would not do. So the thickness of the border K is the only candidate.

3) Incompatibility. If G is both of the form $]-\infty,a] \cup a+K_1$ and of the form $]-\infty,b] \setminus b+K_2$ then $K_1 = K_2 = K$ by unicity. But then K would be both a generalized galaxy and a generalized halo which contradicts Theorem 4.17.

Finally, we remark that the set K has the same complexity as the set G or H, being defined with the aid of the same index set X.

5. External equations.

Definition 4.38: The problem to determine the normal form of a convex subset of an ordered group will be called an external equation. The solution of an external equation is defined to be the normal form in question.

Examples: 1) The normal form of the set $\{x \in \mathbb{R} \mid e^x \simeq \sum_{n=0}^{\omega} \frac{x^n}{n!}\}$ is $([-\omega',\omega']\setminus(-\omega'+G))\setminus(\omega'+G)$ where $\omega' = \frac{\omega}{e} + \frac{\log \omega}{2e}$ (see Chapter I).

2) The normal form of $\{\omega \in \mathbb{N} \mid |R_\omega(\varepsilon)/R_n(\varepsilon)| \underset{\simeq}{<} 1$ for all $n \in \mathbb{N}\}$, where $R_n(\varepsilon) = {}_0\int^\infty e^{-t} f(\varepsilon t)dt - \sum_{k=0}^{n} c_k k!\varepsilon^k$, with $f(t) = \sum_{k=0}^{\infty} c_k t^k$ and $c_{n+1}/c_n = -1+o(1/\sqrt{n})$ for $n \to \infty$, is $(1/\varepsilon + 1/\sqrt{\varepsilon}\text{-hal}) \cap \mathbb{N}$ (see Chapter III; it concerns the set of indices of nearly minimal remainders).

3) In the previous section we solved external equations of the form $F = \{t \in \mathbb{R} \mid x^- \lessapprox f(t) \lessapprox x^+\}$, appearing in the problem of stretching jumps of solutions of differential equations.

4) In Chapter V we present a fairly general method for the asymptotic evaluation of integrals. The method includes the solution of an external equation, in order to locate the domain of the main contribution to the value of the integral.

5) In Chapter VI the set of all real numbers having some prescribed shadow expansion will be brought in normal form.

6) Gibbs phenomenon: The function

$$f(x) = \begin{cases} 1 & 0 < x < \pi \\ 0 & x = 0 \\ -1 & -\pi < x < 0 \end{cases}$$

has the Fourier-series

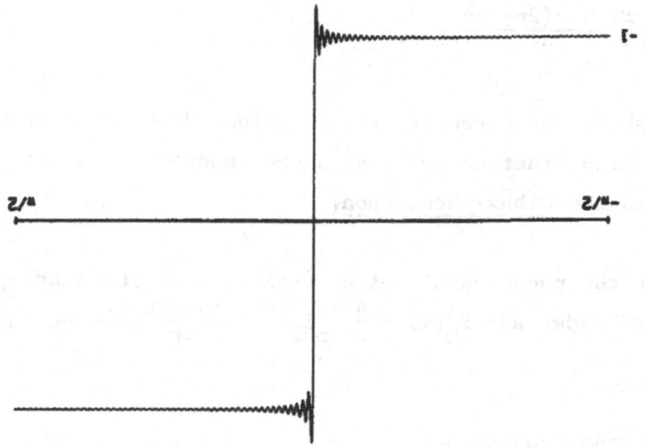

<u>Figure 1</u>: <u>The Gibbs phenomenon</u>: The function

$$f(x) = \begin{cases} 1 & 2k\pi < x < (2k+1)\pi \\ 0 & x = k\pi \\ -1 & (2k+1)\pi < x < (2k+2)\pi \end{cases} \qquad k \in \mathbb{Z}$$

has the Fourier series $\frac{4}{\pi}\sum_{n=0}^{\infty}\frac{\sin(2n+1)x}{2n+1}$. Though this series converges everywhere to the function f, near the discontinuities a trigonometric polynomial of high order does not approximate the function, but oscillates around it with non-negligible amplitude. This is the <u>Gibbs phenomenon</u>. Theoretically, the oscillations of a trigonometric polynomial of unlimited degree ω have appreciable amplitude in the 1/ω-galaxy of the points of discontinuity. The figure shows the graph of $\frac{4}{\pi}\sum_{n=0}^{100}\frac{\sin(2n+1)x}{2n+1}$.

$$f(x) = \frac{4}{\pi} \sum_{n=0}^{\infty} \frac{\sin(2n+1)x}{2n+1}$$

The series is everywhere convergent to the function. However, near the discontinuities the partial sums of high order do not give a good approximation of the function (see Fig. 1). This is the Gibbs phenomenon.

What is the range of the phenomenon? Let us treat the question asymptotically. Let ω be an unlimited odd index and $S_\omega(x) = \frac{4}{\pi} \sum_{n=0}^{(\omega-1)/2} \frac{\sin(2n+1)x}{2n+1}$ be the partial sum of degree ω.

Let us solve the external equation

$$F = {}^{c}\{x \in]0,\tfrac{\pi}{2}[\ \mid S_\omega(x) \not\simeq 1\}$$

(${}^{c}A$ means the convex closure of the set A). We evaluate $S_\omega(x)$ in three regions: (i) $x \in 1/\omega$-gal, (ii) $0 \lneqq x \leqq \pi/2$, (iii) $x \simeq 0$, but not $x \in 1/\omega$-gal.

(i). $x \in 1/\omega$-gal. Put $x = u/\omega$, with limited u. Then by the nonstandard definition of the Riemann integral

$$\frac{4}{\pi} \sum_{n=0}^{(\omega-1)/2} \frac{\sin((2n+1)u/\omega)}{2n+1} = \frac{2}{\pi} \sum_{n=0}^{(\omega-1)/2} \frac{\sin((2n+1)u/\omega)}{(2n+1)u/\omega} 2u/\omega \simeq \frac{2}{\pi} \int_0^u \frac{\sin t}{t} dt$$

(ii). $0 \lneqq x \leqq \pi/2$. One has $S_\omega \simeq 1$, for the Fourier series is uniformly convergent on every interval $[d,\pi/2]$ with $d > 0$.

(iii). $x \simeq 0$, not $x \in 1/\omega$-gal. For all limited a one has

$$S_\omega(x) = \frac{2}{\pi} \sum_{n=0}^{[a/x]-1} \frac{\sin(2n+1)x}{(2n+1)x} .2x + \frac{4}{\pi} \sum_{n=[a/x]}^{(\omega-1)/2} \frac{\sin(2n+1)x}{(2n+1)}$$

$$\simeq \frac{2}{\pi} \int_0^a \frac{\sin t}{t} dt + \frac{4}{\pi} \sum_{n=[a/x]}^{(\omega-1)/2} \frac{\sin(2n+1)x}{2n+1}$$

By Robinson's lemma the above formula still holds for some unlimited α. We may assume that α/x is an integer. Notice that $\frac{2}{\pi} \int_0^\alpha \frac{\sin t}{t} dt \simeq 1$. We will show that $\sum_{n=\alpha/x}^{(\omega-1)/2} \frac{\sin(2n+1)x}{2n+1} \simeq 0$. Indeed, applying Abel's partial summation formula

$$\sum_{n=\alpha/x}^{(\omega-1)/2} \frac{\sin(2n+1)x}{2n+1} = \sum_{n=\alpha/x}^{(\omega-3)/2} \left(\sum_{k=\alpha/x}^{n} \sin(2k+1)x\right)\left(\frac{1}{2n+1}-\frac{1}{2n+3}\right) + \sum_{k=\alpha/x}^{(\omega-1)/2} \frac{\sin(2k+1)\ x}{\omega}$$

$$= \sum_{n=\alpha/x}^{(\omega-3)/2} \frac{\cos(2n+1)x-\cos(2\alpha+x)}{2x}\cdot\frac{2}{(2n+3)(2n+1)} + \frac{\cos\ \omega x-\cos(2\alpha+x)}{2\omega x}$$

Now $\frac{\cos\ \omega x-\cos(2\alpha+x)}{2\omega x} \simeq 0$ and furthermore

$$\left|\sum_{n=\alpha/x}^{(\omega-3)/2} \frac{\cos(2n+1)x-\cos(2\alpha+x)}{2x}\cdot\frac{2}{(2n+3)(2n+1)}\right| \leq \frac{1}{2x} \sum_{n=\alpha/x}^{\omega/2} \frac{1}{n^2} \leq \frac{1}{x} \int_{\alpha/x}^{\omega/2} \frac{1}{t^2} dt \simeq 0$$

Hence $S_\omega \simeq 1$.

We conclude from (i), (ii) and (iii) that $F = 1/\omega\text{-gal}\setminus\{0\}$. So asymptotically the Gibbs phenomenon appears for numbers of the order $1/\omega$ where ω is the inverse of the degree of the trigonometrical polynomial. Notice that the shadow of the function $S_\omega(u/\omega)$ does not depend on ω.

From a nonstandard point of view, the Gibbs phenomenon has also been treated by Laugwitz [32].

7) The term "external equation" is borrowed from T. Ismael who in [30] considers the problem of concentration of curvature $C(t)$ of parametrized curves $(x(t),y(t))$. In particular he studies external equations of the form $\{t \mid C(t) \gtrapprox 0\}$.

As illustrated by the above examples, many external equations can be reduced to one of the forms

$$G = \{x \in \mathbb{R} \mid f(\omega,x) \in G\}$$
$$H = \{x \in \mathbb{R} \mid f(\omega,x) \in \text{hal}(0)\}$$

where $f: \mathbb{R} \times \mathbb{R} \to \mathbb{R}$ is a standard function, which may be thought to be increasing in x and continuous, and ω is some unlimited number. For such equations, we will strengthen the definition of a solution, i.e. we will give further precision to the notion "normal form". First we make some simplifying remarks.

Without loss of generality, we will from now on only consider external equations of the first form, and suppose that $f(\omega,.)$ is bounded below by some limited number . To avoid the trivial internal case, we suppose in addition that $f(\omega,.)$ takes both limited and unlimited values. Then G is a lower halfline, and a galaxy.

<u>Definition 4.39</u>: Let $G = \{x \in \mathbb{R} \mid f(\omega,x) \in G\}$ where $f: \mathbb{R} \times \mathbb{R} \to \mathbb{R}$ is a standard function, $\omega \in \mathbb{R}$ is unlimited, and $f(\omega,x)$ is increasing and continuous in x, taking both positive limited and unlimited values, and bounded below by some limited number.

Then a strong normal form of G is either a representation $G =]-\infty,a(\omega)] \cup a(\omega)+K$, where $a: \mathbb{R} \to \mathbb{R}$ is standard and K is a convex additive group, which is written in the form $K = \bigcup_{n\in\mathbb{N}} [-\varphi(\omega,n),\varphi(\omega,n)]$, where $\varphi: \mathbb{R} \times \mathbb{N} \to \mathbb{R}$ is standard, or a representation $G =]-\infty,\overline{b}(\omega)] \setminus b(\omega)+L$, where $b: \mathbb{R} \to \mathbb{R}$ is standard and L is a convex additive group, which is written in the form $L = \bigcap_{n\in\mathbb{N}} [-\Psi(\omega,n),\Psi(\omega,n)]$, where $\Psi: \mathbb{R} \times \mathbb{N} \to \mathbb{R}$ is standard.

Theorem 4.40: Every external equation $G = \{x \in \mathbb{R} \mid f(\omega,x) \in \mathbb{G}\}$, where $f: \mathbb{R} \times \mathbb{R} \to \mathbb{R}$ is standard, $\omega \in \mathbb{R}$ is unlimited and $f(\omega,x)$ is increasing and continuous in x, taking both limited and unlimited values and bounded below by some limited number has a solution in strong normal form.

Proof: The galaxy G is a lower halfline of $\mathbb{R}$. Let H be its complement in $\mathbb{R}$. Then H is a halo and an upper halfline.

Define a standard (partial) function $f: \mathbb{R} \times \mathbb{N} \to \mathbb{R}$ by

$$g(y,n) = \inf\{x \in \mathbb{R} \mid n = f(y,x)\}$$

Then $g(\omega,n)$ is defined at least for all st n larger than some st $\overline{n}$ and $G = \bigcup_{n\in\mathbb{N}}]-\infty,g(\omega,n)]$.

From now on we imitate parts of the proof of Theorem 4.34. Firstly, if G is of the form $]-\infty,\overline{a}] \cup \overline{a}+K$, then there is some standard $\tilde{n}$ such that $g(\omega,\tilde{n}) > \overline{a}$. Define standard functions $a: \mathbb{R} \to \mathbb{R}$ and $\varphi: \mathbb{R} \times \mathbb{N} \to \mathbb{R}$ by

$$a(y) = g(y,\tilde{n})$$

$$\varphi(y,n) = \begin{cases} g(y,n)-g(y,\tilde{n}) & n \geq \tilde{n} \\ 0 & \text{otherwise} \end{cases}$$

Then $K = \bigcup_{n\in\mathbb{N}} [-\varphi(\omega,n),\varphi(\omega,n)]$ and $G =]-\infty,a(\omega)] \cup a(\omega)+K$.

Secondly, suppose G is of the form $]-\infty,b] \setminus b+K$. We define a standard sequence $(n_k)_{k\in\mathbb{N}}$ by external induction. Put $n_0 = \overline{n}$. If $n_0,\dots,n_{k-1}$ are defined, put $n_k = \min\{n \mid g(\omega,n)-g(\omega,n_{k-1}) \in \Delta\}$. Let $(n_k)_{k\in\mathbb{N}}$ be the standardized of $(n_k)_{k\in\mathbb{N}}$. Define the standard function $\Psi: \mathbb{R} \times \mathbb{N} \to \mathbb{R}$ by $\Psi(y,k) = g(y,n_{k+1})-g(y,n_k)$. Then $K = \bigcap_{k\in\mathbb{N}} [-\Psi(\omega,k),\Psi(\omega,k)]$.

Finally, we define a standard function $b: \mathbb{R} \to \mathbb{R}$ such that $G =]-\infty,b(\omega)] \setminus b(\omega)+K$. Let $(n_p)_{p\in\mathbb{N}}$ be the standardized of the sequence $(n_p)_{p\in\mathbb{N}}$ of standard numbers defined by $n_p = \min\{n \mid g(\omega,n)+\Psi(\omega,p)\} \in H$. Let $\overline{g}(y,p)$ be the standard function $g(y,n_p)$ and $\overline{h}(y,p)$ be the standard function $\overline{g}(y,p)+\Psi(y,p)$. Except possibly for two standard permutations of $\mathbb{N}$, we may assume that $\overline{g}(\omega,p)$ is increasing in p and $\overline{h}(\omega,p)$ is decreasing in p.

Now there are two possibilities depending on $y \in \mathbb{R}^+$

1) $(\overline{g}(y,p))_{p\in\mathbb{N}}$ is increasing and $\overline{h}(y,p)_{p\in\mathbb{N}}$ is decreasing for all $p \in \mathbb{N}$. Then put

$$b(y) = \begin{cases} \overline{g}(y,[y]) & \text{if } \overline{g}(y(p)) < \overline{h}(y,p) \text{ for all } p \in \mathbb{N} \\ \overline{g}(y,\overline{p}(y)) & \text{if } \overline{p}(y) \text{ is minimal such that } \overline{g}(y,p+1) \geqq \overline{h}(y,p+1) \end{cases}$$

2) If not, then there is a minimal $q(y)$ such that $\overline{g}(y,q(y)+1) \leqq \overline{g}(y,q(y))$ or $\overline{h}(y,q(y)+1) \geqq \overline{h}(y,q(y))$. Then put

$$b(y) = \begin{cases} \overline{g}(y,q(y)) & \text{if } \overline{g}(y,q(y)) < \overline{h}(y,q(y)) \\ \overline{g}(y,\overline{p}(y)) & \text{if } \overline{p}(y) \text{ is minimal such that } \overline{g}(y,p+1) \geqq \overline{h}(y,p+1) \end{cases}$$

All ingredients in the definition of the function b are standard, so b is standard. Moreover, in all four cases one has $\overline{g}(\omega,p) < b(\omega) < \overline{h}(\omega,p)$ for all st $p \in \mathbb{N}$. Hence $G =]-\infty,b(\omega)] \smallsetminus b(\omega)+K$.

CHAPTER V. APPROXIMATION LEMMA'S.

Now, we have to discuss and justify the tools we used in the preceding chapters. We present some lemma's of general impact, they unify a certain number of procedures to get approximations of numbers and functions. Notice that several lemma's are related to classical theorems (uniform convergence, Lebesgue's theorem on dominated convergence, method of comparing sums and integrals). However, these theorems are usually applied in a different context: existence of limits, convergence, majoration,... . The chapter will be devided in two parts: approximation of integrals and approximation of sums.

1. Asymptotic approximations of integrals.

We consider real and Riemann-integrable functions. The first lemma has already been stated by many authors.

Lemma 5.1: Let f and g be internal functions such that $f(x) \simeq g(x)$ for all elements x of an interval $[a,b]$ of limited length. Then

$$\int_a^b f(x)dx \simeq \int_a^b g(x)dx$$

Proof: We define the function $h: [a,b] \to \mathbb{R}$ by $h(x) = f(x)-g(x)$. Then $h(x) \simeq 0$ for all $x \in [a,b]$. Let $S = \sup_{x\in[a,b]} |h(x)|$. Then $S \simeq 0$. So

$$\left|\int_a^b f(x)dx-\int_a^b g(x)dx\right| \leqq \int_a^b |h(x)|dx \leqq S(b-a) \simeq 0$$

In the next lemma we state a condition which guarantees that the near-equality in Lemma 5.1 still holds for integrals over unlimited or unbounded intervals. The lemma is related to Lebesgue's dominated convergence theorem and the Weierstrass M-test of uniform convergence.

Lemma 5.2: (Lemma of dominated approximation). Let f and g be two internal functions such that $f(x) \simeq g(x)$ for all limited x. Let h be a standard integrable function such that $|f(x)|,|g(x)| \leqq h(x)$ for all $x \in \mathbb{R}$. Then

$$\int_{-\infty}^{\infty} f(x)dx \simeq \int_{-\infty}^{\infty} g(x)dx$$

Proof: By Lemma 5.1 one has for all limited $n \in \mathbb{N}$

$$\int_{-n}^{n} f(x)dx \simeq \int_{-n}^{n} g(x)dx$$

By Robinson's lemma, there exists some unlimited $\omega \in \mathbb{N}$ such that

$$\int_{-\omega}^{\omega} f(x)dx \simeq \int_{-\omega}^{\omega} g(x)dx$$

Now $\int_{|x|\geq\omega} h(x)dx \simeq 0$, for h is standard integrable. So $\int_{|x|\geq\omega} f(x)dx \simeq 0$ and $\int_{|x|\geq\omega} g(x)dx \simeq 0$. Hence $\int_{-\omega}^{\omega} f(x)dx \simeq \int_{-\infty}^{\infty} f(x)dx$ and $\int_{-\omega}^{\omega} g(x)dx \simeq \int_{-\infty}^{\infty} g(x)dx$. One concludes that

$$\int_{-\infty}^{\infty} f(x)dx \simeq \int_{-\infty}^{\infty} g(x)dx$$

With the above lemma's we may approximate the integral of a function close to a function whose primitive is known. Note that the majoration h of lemma 5.2 is used only in the justification of the approximation, and does not affect the quality of the approximation. Hence in practice the choice of the majoration is led by the pure reason of convenience.

Lemma 5.2 may be adapted to approximations of functions on the other topological galaxies, for instance:

Lemma 5.3: Let f and g be two internal functions such that $f(x) \simeq g(x)$ for all $x \in \mathbb{A}^+$. Let h be a standard integrable function such that $|f(x)|, |g(x)| \leqq h(x)$ for all $x > 0$. Then $\int_0^{\infty} f(x)dx \simeq \int_0^{\infty} g(x)dx$.

The lemma of dominated approximation cannot be used if $\{x \mid f(x) \not\simeq 0\}$ is too big: because the majoration h is standard integrable one needs $\{x \mid f(x) \not\simeq 0\} \subset \mathbb{C}$. On the other hand, the set $\{x \mid f(x) \not\simeq 0\}$ may not be too small. Suppose, for instance, that $\{x \mid f(x) \not\simeq 0\} \subset \text{hal}(a)$ for some a. Then the lemma of dominated approximation can only give $\int_{-\infty}^{\infty} f(x)dx \simeq 0$, an approximation which is correct, but not very useful.

Let us define a class of functions, whose integrals are particularly suited to approximations based on the lemma of dominated approximation.

Definition 5.4: A function f is said to be of class A if (i) f is limited, (ii) $\{x \mid f(x) \not\simeq 0\} \subset \mathbb{C}$, (iii) there exists $a \in G$ such that $\{x \mid f(x) \not\simeq 0\} \supset \text{hal}(a)$.

Let f be a function of class A, let g be a function such that $f(x) \simeq g(x)$ on $\mathbb{C}$, and suppose h is a standard integrable majoration of $|f|$ and $|g|$. Then at least one of the approximations $\int_{-\infty}^{\infty} f^+(x)dx \simeq \int_{-\infty}^{\infty} g^+(x)dx$ or $\int_{-\infty}^{\infty} f^-(x)dx \simeq \int_{-\infty}^{\infty} g^-(x)dx$ is significative, being the near-equality of two appreciable numbers.

How to transform a more general integrand into an integrand of class A?

Next example illustrates a strategy which can be used to "focus" such integrands.

Example 5.5: (related to proposition 1.6 of Chapter I). Let $t > 0$, let p be limited and ω be positive unlimited. Define $f(v)$ and J by

$$f(v) = \frac{(t-v)^{\omega-1}}{(1+v)^{\omega+p}} \qquad (0 \leqq v \leqq t)$$

$$J = {}_0\int^t f(v)dv$$

The integrand takes it maximum for $v = 0$. Multiplying and dividing by the maximum value $f(0)$ we get

$$J = t^{\omega-1} {}_0\int^t \frac{(1-v/t)^{\omega-1}}{(1+v)^{\omega+p}}dv \equiv t^{\omega-1} {}_0\int^t g(v)dv$$

The new integrand is everywhere limited. We now solve the external equation $F = \{v \mid g(v) \not\gtrsim 0\}$. Applying Euler's formula $(1+y/\omega)^{\omega} \simeq e^y$ we find (i) $F = (t/\omega\text{-gal})^+$ if t is limited and (ii) $F = (1/\omega\text{-gal})^+$ if t is unlimited. In case (i) we obtain an integrand of class A by the substitution $v = \frac{tu}{\omega}$. Then

$$J = \frac{t^{\omega}}{\omega} {}_0\int^{\omega} \frac{(1-u/\omega)^{\omega-1}}{(1+tu/\omega)^{\omega+p}}du$$

Put $I(u) = (1-u/\omega)^{\omega-1}/(1+tu/\omega)^{\omega+p}$. Then $I(u) \simeq e^{-(1+t)u}$ for all limited u and $I(u) \leqq (1-u/\omega)^{\omega-1} \leqq e^{-u/2}$ for all $u \leqq \omega$. So by the lemma of dominated approximation ${}_0\int^{\omega} I(u)du \simeq {}_0\int^{\infty} e^{-(1+t)u}du \simeq \frac{1}{1+t}$. Hence

$$J = (1+\emptyset)\frac{t^{\omega}}{\omega(1+t)}$$

In case (ii) we obtain an integrand of class A by the substitution $v = s/\omega$. Then

$$J = \frac{t^{\omega-1}}{\omega} {}_0\int^{\omega t} \frac{(1-s/(t\omega))^{\omega-1}}{(1+s/\omega)^{\omega+p}} ds$$

Again with the lemma of dominated approximation, we find ${}_0\int^{\omega t} \frac{(1-s/(t\omega))^{\omega-1}}{(1+s/\omega)^{\omega+p}}ds \simeq {}_0\int^{\infty} e^{-s}ds = 1$. Then

$$J = (1+\emptyset)\frac{t^{\omega-1}}{\omega}$$

Let us make two remarks: (1) of course for $t \in A^+$ we have the choice between both substitutions, i.e. $v = tu/\omega$ and $v = u/\omega$, (2) We found in case (ii) the same first order approximation as in case (i), for $t^{\omega-1} = (1+\emptyset)\frac{t^{\omega}}{1+t}$.

Essential steps in the above argument were:

1) We devided $f(v)$ by $\max_{v \geq 0} f(v)$ and thus got an everywhere limited function $g(v)$ taking appreciable values at least somewhere.

2) We solved the external equation $F = \{v \mid g(v) \not\simeq 0\}$; its solution suggested a substitution to obtain an integrand of class A. The substitution was a bijection of $\mathbb{G}$ onto F.

This method may be applied to the problem of focusing bounded functions $f(x)$ such that $\lim_{x \to \infty} f(x) = 0$, and having a standard number of extrema. To fix ideas, assume such a function to be positive, continuous, and having only one extremum, which is a maximum. Then

Step 1: Localize the maximum and translate it to 0. Devide the function by its maximum value. The new integrand, say, $g(t)$ is now limited everywhere, and has its maximum (1) at 0.

We discuss $\int_0^{\infty} g(t)dt$. The left integrand $\int_{-\infty}^{0} g(t)dt$ may be handled analogously.

Step 2: Solve the external equation $F = \{t \mid g(t) \gtrsim 0\}$. By proposition 4.26(ii), there exists a C^{∞} bijection, say B, mapping $\mathbb{G}^{+}$ onto F. Put $t = B(s)$. Then the function $g(B(s))$ is of class A. Now the integral reads $\int_0^{\infty} g(B(s))B'(s)ds$.

Step 3: Case (1): If B could be taken linear, i.e. if $F = (\alpha\text{-gal})^{+}$ for some $\alpha \in \mathbb{R}$, then $B'(s)$ is just a number, which can be removed from under the integral sign. The resulting integrand is a function of class A. Try to apply the lemma of dominated approximation. In the bad case $g(B(s))$ does not have a standard majoration.

Case (2): If F is not the positive part of an α-galaxy, then one could start the procedure over again with $g(B(s))B'(s)$. But it would be advisable to look for an ad hoc substitution.

Of course, these steps are only guidelines and one should not refuse any shortcut. But this method helped me to face some unknown situations.

Here is an example where the method cannot be applied. Put $f(x) = \frac{1}{\log x + \varepsilon x^2}$ for $x \geq 2$, where $\varepsilon \simeq 0$. For limited x one has $f(x) \simeq \frac{1}{\log x}$. But $\int_2^{\infty} f(x)dx \not\simeq \int_2^{\infty} \frac{dx}{\log x}$, because the first integral is convergent and the second is divergent.

We now consider integrals of the form $_0\int^\infty e^{-\omega\varphi(t)}\Psi(t)dt$, where φ is some standard increasing function, and Ψ is, say, near-standard. Next lemma shows to which extent only the values of the integrand near 0 contribute in a significant way to the value of the integral.

Lemma 5.6 (Concentration lemma): Let φ be a standard function, defined on $[0,\infty[$, increasing, having an isolated minimum in 0, and such that $\varphi(t) \geqq mt^r$ $(m,r > 0)$ beyond some $y \geqq 0$. Let Ψ be an internal function, defined on $[0,\infty[$, such that for all $d \gneqq 0$ there exist standard constants K and C such that $|\Psi(t)| \leqq Ke^{\varphi(t)}$ for all $t \geqq d$. Let ω be positive and unlimited. Then there exists $\delta \simeq 0$ such that

$$_\delta\int^\infty e^{-\omega\varphi(t)}\Psi(t)dt = \mu e^{-\omega\varphi(0)} \qquad (\mu \in 1/\omega\text{-M})$$

Proof: Let $d \gneqq 0$. Write $\lambda = e^{-\omega(\varphi(0)-\varphi(d))}$. Because $\varphi(d) \gneqq \varphi(0)$, one has $\lambda \in 1/\omega\text{-M}$. Let st y be such that $\varphi(t)-\varphi(d) \geqq mt^r$ for all $t \geqq y$. Then

$$|_d\int^\infty e^{-\omega\varphi(t)}\Psi(t)dt| \leqq e^{-\omega\varphi(d)}Ke^{C\varphi(d)} \quad _d\int^\infty e^{-\omega(1-C/\omega)(\varphi(t)-\varphi(d))}dt$$

$$\leqq e^{-\omega\varphi(0)}.\lambda.Ke^{C\varphi(d)}(_d\int^y e^{-\omega(1-C/\omega)(\varphi(t)-\varphi(d))}dt$$

$$+ {_y\int^\infty} e^{-\omega(1-C/\omega)mt^r}dt$$

$$\leqq e^{-\omega\varphi(0)}\lambda.Ke^{C\varphi(d)}(y-d+1)$$

$$= \mu e^{-\omega\varphi(0)} \qquad (\mu \in 1/\omega\text{-M})$$

Hence $e^{\omega\varphi(0)}{_d\int^\infty} e^{-\omega\varphi(t)}\Psi(t)dt$ belongs to the $1/\omega$-microhalo for all $d \gneqq 0$. By the Fehrele principle, there exists $\delta \simeq 0$ such that $e^{\omega\varphi(0)}{_\delta\int^\infty} e^{-\omega\varphi(t)}\Psi(t)dt \in 1/\omega\text{-M}$.

Comment: Lemma 5.6 shows that an integral of type $J(\omega) = {_0\int^\infty} e^{-\omega\varphi(t)}\Psi(t)dt$ can very closely be approximated by an integral $_0\int^\delta e^{-\omega\varphi(t)}\Psi(t)dt$ where δ is a sufficiently large infinitesimal. The contraction of the interval of integration is notably useful if φ and Ψ can be approximated. Typical approximations are $\varphi(t) = t^r(1+\emptyset)$ and $\Psi(t) \simeq \Psi(0)$ for $t \simeq 0$. Then the approximation will be valid on the whole interval of integration, yielding easily a first approximation of $J(\omega)$.

Suppose one wishes to develop $J(\omega)$ in powers of $1/\omega$. As far as standard powers are concerned, it again suffices to integrate over an interval of infinitesimal length, for the error made in substituting the infinite interval $[0,\infty[$ by the interval $[0,\delta]$ can be supposed to be smaller than all standard powers of $1/\omega$.

The concentration lemma will be used in the proof of the next lemma, which is a nonstandard version of the classical "method of Laplace".

Lemma 5.7: Let φ be a standard function, defined and increasing on $[0,\infty[$, such that $\varphi(t) = (1+\emptyset)at^r$ for $t \simeq 0$ (st $a,r > 0$) and $\varphi(t) \geqq mt^q$ beyond some $y \geqq 0$ ($m,q > 0$). Let Ψ be an internal function, defined on $]0,\infty[$ such that $\Psi(t) = (1+\emptyset)bt^s$ for $t \simeq 0$ (st $b \neq 0$, st $s > -1$) and such that for all $d \gtrapprox 0$ there exist standard constants K and C such that $|\Psi(t)| \leqq Ke^{C\varphi(t)}$ for all $t \geqq d$. Let Ψ be positive and unlimited. Then

$$_0\int^\infty e^{-\omega\varphi(t)}\Psi(t)dt = (1+\emptyset)\frac{b\Gamma(\frac{s+1}{r})}{ra^{\frac{s+1}{r}}}\cdot\frac{1}{\omega^{\frac{s+1}{r}}}$$

Proof: By the concentration lemma, let $\delta \simeq 0$ be such that $_\delta\int^\infty e^{-\omega\varphi(t)}\Psi(t)dt \in 1/\omega\text{-}M$. We may suppose that $\delta > 1/\omega^{1/n}$ for all st $n \in \mathbb{N}$. Then

$$_0\int^\infty e^{-\omega\varphi(t)}\Psi(t)dt = {_0\int^\delta} e^{-(1+\emptyset)\omega at^r}(1+\emptyset)bt^s dt+\mu \qquad (\mu \in 1/\omega\text{-}M)$$

The solution of the external equation $F = \{t \mid e^{-(1+\emptyset)\omega at^r}(1+\emptyset)bt^s \not\simeq 0\}$ is $(1/\omega^{1/r}\text{-gal})^+$. Hence we obtain an integrand of class A by the substitution $t = (\frac{u}{a\omega})^{1/r}$. Then

$$_0\int^\infty e^{-\omega\varphi(t)}\psi(t)dt = \frac{b}{ra^{\frac{s+1}{r}}}\cdot\frac{1}{\omega^{\frac{s+1}{r}}}\ {_0\int^{a\omega\delta^r}} e^{-(1+\emptyset)u}(1+\emptyset)u^{\frac{s+1}{r}-1}du+\mu$$

Put $I(u) = e^{-(1+\emptyset)u}(1+\emptyset)u^{\frac{s+1}{r}-1}$. Then $I(u) \simeq e^{-u}\ u^{\frac{s+1}{r}-1}$ for all appreciable u and $I(u) \leqq 2e^{-u/2}\ u^{\frac{s+1}{r}-1}$ for all $u \leqq a\omega\delta^r$. Then by lemma 5.3

$$_0\int^\infty I(u)du \simeq {_0\int^\infty} e^{-u}\ u^{\frac{s+1}{r}-1}du = \Gamma(\tfrac{s+1}{r})$$

Consequently $_0\int^\infty e^{-\omega\varphi(t)}\psi(t)dt = (1+\emptyset)\frac{b\Gamma(\frac{s+1}{r})}{ra^{\frac{s+1}{r}}}\cdot\frac{1}{\omega^{\frac{s+1}{r}}}$

Remark: The above lemma also holds for integrals $_0\int^c e^{-\omega\varphi(t)}\Psi(t)dt$ where $c \gtrapprox 0$.

Examples: 1) For $\varphi(t) = t$ the integral is a Laplace transform. If for instance $\psi(t) = (1+\emptyset)bt^n$ near 0 (st $b \not\simeq 0$, st $n \in \mathbb{N}$), and is of S-exponential order, then

$$_0\int^\infty e^{-\omega t}\Psi(t)dt = (1+\emptyset)bn!/\omega^{n+1}$$

2) Suppose φ is two times continuously differentiable in 0, such that $\varphi'(0) = 0$ and $\varphi''(0) > 0$ (so that φ has a minimum in 0). If $\Psi(0) \not\simeq 0$, and Ψ is S-continuous in 0, and satisfies the growth conditions of Lemma 5.7 in both directions, then

$$\int_{-\infty}^{\infty} e^{-\omega\varphi(t)}\Psi(t)dt = (1+\emptyset)\Psi(0)\sqrt{\frac{2\pi}{\varphi''(0)}}\cdot\frac{1}{\sqrt{\omega}}$$

3) Stirling's formula

$$\omega! = \int_0^{\infty} e^{-t}t^{\omega}dt = (1+\emptyset)\omega^{\omega}e^{-\omega}\sqrt{\omega}\sqrt{2\pi} \qquad (\omega \text{ positive unlimited})$$

is a direct consequence of lemma 5.6. Indeed, put $t = \omega(1+u)$. Then

$$\int_0^{\infty} e^{-t}t^{\omega}dt = \omega^{\omega}e^{-\omega}\omega \int_{-1}^{\infty} e^{-\omega(u-\log(1+u))}du$$

Now we are in the case of example 2. Hence $\omega! = (1+\emptyset)\omega^{\omega}e^{-\omega}\sqrt{\omega}\sqrt{2\pi}$.

More in general, the formula of Theorem 1.4

$$\int_0^{\infty} e^{-t}t^{\omega}\Psi(\varepsilon t)dt = (1+\emptyset)\Gamma(\omega+1)\Psi(\varepsilon\omega)$$

which helped us to find the approximation factor for divergent expansions (Chapters II and III), is also a direct consequence of lemma 5.6.

4) The function $\Gamma(1+x)$ has the Taylor expansion

$$\Gamma(1+x) = \sum_{n=0}^{\infty} \frac{\gamma_n}{n!}x^n$$

with $\gamma_n = \int_0^{\infty} (\log t)^n e^{-t}dt$. Because the function $\Gamma(1+x)$ has an isolated singularity in -1 it follows from "Darboux's theorem " (see page 54) that $\gamma_{\omega} = (1+\emptyset)(-1)^{\omega}\omega!$ for unlimited ω. Let us check this directly. We treat $\int_0^1 (\log t)^{\omega}e^{-t}dt$ and $\int_1^{\infty} (\log t)^{\omega}e^{-t}$ separately.

(i) $\int_0^1 (\log t)^{\omega}e^{-t}dt$: Put $t = e^{-u}$. Then

$$\int_0^1 (\log t)^{\omega}e^{-t}dt = (-1)^{\omega} \int_0^{\infty} e^{-u}u^{\omega}e^{-e^{-u}} du.$$

Put $\Psi(u) = e^{-e^{-\omega u}}$. Then Ψ satisfies the conditions of Theorem 1.4 (with an innocent exception: Ψ is not S-continuous at 0). So by Theorem 1.4

$$\int_0^{\infty} e^{-u}u^{\omega}\Psi(u/\omega)du = (1+\emptyset)\omega!\Psi(1) = (1+\emptyset)\omega!$$

Hence $\int_0^1 (\log t)^\omega e^{-t}dt = (1+\emptyset)(-1)^\omega \omega!$

(ii) $\int_1^\infty (\log t)^\omega e^{-t}dt$: Note that $\int_1^\infty (\log t)^\omega e^{-t}dt \leq \int_1^\infty e^{-t}t^{\omega/2}dt = \Gamma(\frac{\omega}{2}+1) = \emptyset.\omega!$

Combining, one sees that $\gamma_\omega = (1+\emptyset)(-1)^\omega \omega!$

2. Asymptotic approximations of sums.

The first two lemma's are elementary, but very useful

Lemma 5.8: Let $(a_k)_{k\in\mathbb{N}}$ and $(b_k)_{k\in\mathbb{N}}$ be internal sequences with strictly positive terms, and let $n \in \mathbb{N}$ be arbitrary. Suppose $b_k = (1+\alpha_k)a_k$ with $\alpha_k \simeq 0$ for all $k \leq n$. Then

$$\sum_{k=0}^{n} b_k = (1+\emptyset)\sum_{k=0}^{n} a_k$$

Lemma 5.9: Let f be an internal real decreasing and positive function such that $f(0) \simeq 0$, and let $n \in \mathbb{N}$ be arbitrary. Then

$$\sum_{k=0}^{n} f(k) \simeq \int_0^n f(x)dx$$

With these two lemma's we can approximate the growth of the terms of standard slowly varying sequences.

Lemma 5.10: Let $(c_n)_{n\in\mathbb{N}}$ be a standard sequence of positive real numbers and $\omega \in \mathbb{N}$ be unlimited. Then

(i) if $\lim_{n\to\infty} c_{n+1}/c_n = 1$, then $c_\omega = e^{\emptyset.\omega}$

(ii) if $c_{n+1}/c_n = 1+a/n^p+o(1/n^p)$ for $n \to \infty$, where $a \in \mathbb{R}$ and $0 < p < 1$, then
$$c_\omega = e^{\frac{(a+\emptyset)}{1-p}\omega^{1-p}}.$$

(iii) If $c_{n+1}/c_n = 1+a/n+o(1/n)$ for $n \to \infty$, where $a \in \mathbb{R}$, then $c_\omega = \omega^{a+\emptyset}$.

Proof: We only prove (ii). Let $\nu \in \mathbb{N}$ be unlimited such that $\nu/\omega \simeq 0$ and $c_\nu = e^{\emptyset.\omega^{1-p}}$. Then

$$c_\omega = \frac{c_\omega}{c_{\omega-1}} \cdots \frac{c_{\nu+1}}{c_\nu}.c_\nu$$

$$= c_\nu e^{\sum_{n=\nu}^{\omega-1} \log(1+(a+\emptyset)/n^p)}$$

$$= c_\nu e^{\sum_{n=\nu}^{\omega-1} (a+\emptyset)/n^p}$$

$$= c_\nu e^{(a+\emptyset)\sum_{n=\nu}^{\omega-1} 1/n^p}$$

$$= c_\nu e^{(a+\emptyset)\int_\nu^\omega x^{-p}dx}$$

$$= e^{\emptyset.\omega^{1-p} + \frac{(a+\emptyset)}{1-p}(\omega^{1-p}-\nu^{1-p})}$$

$$= e^{\frac{a+\emptyset}{1-p}\omega^{1-p}}$$

Next lemma is a series version of the lemma of dominated approximation.

Lemma 5.11: Let $(u_n)_{n\in\mathbb{N}}$ and $(v_n)_{n\in\mathbb{N}}$ be two internal sequences such that $u_n \simeq v_n$ for all st $n \in \mathbb{N}$. Let $\sum_{n=0}^{\infty} w_n$ be a standard convergent series such that $|u_n|, |v_n| \leqq w_n$ for all $n \in \mathbb{N}$. Then

$$\sum_{n=0}^{\infty} u_n \simeq \sum_{n=0}^{\infty} v_n$$

The last lemma concerns series whose terms decrease very rapidly at the beginning. It states that only the first nonzero term contributes in a significative way to its sum. The lemma is almost contained in lemma 5.11. But we state it separately, because of its frequent use in relation to shadow expansions.

Lemma 5.12: Let $(u_n)_{n\in\mathbb{N}}$ be an internal sequence of real numbers. Let $(n_k)_{k\in\mathbb{N}}$ be the sequence of the indices of its nonzero terms. Suppose $u_{n_1}/u_{n_0} \simeq 0$, and $u_{n_{k+1}}/u_{n_k} \lessapprox 1$ for all k. Then

$$\sum_{n=0}^{\infty} u_n = (1+\emptyset)u_{n_0}$$

p	$\log c_{10000}$	$\frac{10000^{1-p}}{1-p}$
0.1	3656.56	4423.41
0.2	1798.73	1981.12
0.3	856.01	901.37
0.4	405.17	418.65
0.5	194.25	200.00
0.6	95.59	99.53
0.7	48.86	52.83
0.8	26.26	31.55
0.9	15.01	25.12

Figure 1: If $(c_n)_{n\in\mathbb{N}}$ is a standard sequence such that $\frac{c_{n+1}}{c_n} = 1 + \frac{a}{n^p} + o(\frac{1}{n^p})$ for $n \to \infty$, where $0 < p < 1$, we know approximately the growth of the c_n: for unlimited ω

$$\log c_\omega = \frac{(a+\emptyset)}{1-p}\omega^{1-p}$$

In the table above we took $c_n = \prod_{k=1}^{n-1} (1 + \frac{1}{k^p})$, such that $a = 1$, and $\omega = 10000$. The exponent p has the values $p = 0.1,\ldots,0.9$. If $p = 1$, i.e. $\frac{c_{n+1}}{c_n} = 1+\frac{a}{n}+o(\frac{1}{n})$ for $n \to \infty$ then we have asymptotically

$$c_\omega = \omega^{a+\emptyset}$$

For a numerical example of this case (concerning $c_n = \frac{\Gamma(n+a)}{\Gamma(n)}$) see Fig. 16 of Chapter I.

CHAPTER VI. SHADOW EXPANSIONS.

In 1886 Poincaré presented his well-known definition of asymptotic expansions of functions (see definition 2.4 of this book). Poincaré's definition turned out to be highly operational and very practical, as is evident from the literature. However, as regards to pointwise approximation, the definition has some drawbacks: if, say, $\sum_{n=0} c_n x^n$ is the asymptotic expansion of some function f, then no account is given of how a number $f(\overline{x})$ is approximated by the partial sums $\sum_{n=0}^{k} c_n \overline{x}^n$. Also, the definition does not reflect a feature judged characteristic by many authors (e.g. [43], [54]): if $\sum_{n=0} c_n x^n$ is divergent, then for numbers $\overline{x}$ where the series is used for approximation, the terms $c_n \overline{x}^n$ first decrease towards some minimum, then start to grow, and eventually take very high values. As far as approximation is concerned, it is clear that the process of summation may not be continued indefinitely, but this warning is not contained in Poincaré's definition, which is given for all $n \in \mathbb{N}$.

In [18] F. Diener introduces a sort of asymptotic expansions of numbers: the expansions in ε-shadow. This definition does not have the above mentioned drawbacks. Essentially, the definition concerns only standard indices. Furthermore, the definition shows very clearly that for these indices the terms decrease, and that their sum gives better and better approximations. For nonstandard indices the expansion is continued formally, so it is not surprising that sometimes any relation between its partial sums and the original number is lost. Mrs. Diener's definition has another advantage: it is close to a "natural" expansion, namely Goze's decomposition of infinitesimal vectors (this decomposition is "natural" in the sense that there are no preliminary conditions to its existence).

In this chapter we first consider the Goze decomposition (Section 1). In Section 2 we introduce "order scales". In Section 3, Mrs. Diener's definition is extended to shadow expansions with respect to arbitrary order scales. Furthermore, elementary fundamental properties of shadow expansions are considered: uniqueness, existence theorem, relation to the Goze decomposition. We also consider shadow expansions of functions. In Section 4 we discuss the approximation properties of shadow expansions. In the last section we reconsider the problem of the "summation to the index of the smallest term".

1. The Goze decomposition.

Theorem 6.1: (Goze [27]): Let $\binom{\varepsilon_1}{\eta_1}$ be a vector in $\mathbb{R}^2$ with $\varepsilon_1 \simeq 0$ and $\eta_1 \simeq 0$. Then there exist two linear independent standard vectors V_1 and V_2, and two infinitesimals ε_2, η_2 such that

$$\binom{\varepsilon_1}{\eta_1} = \varepsilon_2 V_1 + \varepsilon_2 \eta_2 V_2$$

Proof: We distinct two cases: (i) η_1/ε_1 limited, (ii) η_1/ε_1 unlimited.

(i) η_1/ε_1 limited: Let $a = {}^{o}(\eta_1/\varepsilon_1)$. One has $\eta_1 = a\varepsilon_1+\eta_2\varepsilon_1$ with $\eta_2 \simeq 0$. Put $\varepsilon_2 = \varepsilon_1$. Then

$$\binom{\varepsilon_1}{\eta_1} = \varepsilon_2\binom{1}{a} + \varepsilon_2\eta_2\binom{0}{1}$$

The standard vectors $V_1 = \binom{1}{a}$ and $V_2 = \binom{0}{1}$ are independent.

(ii) η_1/ε_1 unlimited: Then $\eta_1/\varepsilon_1 \simeq 0$. Put $\varepsilon_2 = \eta_1$ and $\eta_2 = \varepsilon_1/\eta_1$. Then

$$\binom{\varepsilon_1}{\eta_1} = \varepsilon_2\binom{0}{1} + \varepsilon_2\eta_2\binom{1}{0}$$

Now define $V_1 = \binom{0}{1}$ and $V_2 = \binom{1}{0}$. Again V_1 and V_2 are standard independent vectors.

Comments: 1) The above finite development of an infinitesimal vector is called a Goze-decomposition, after M. Goze who introduced the development in [27]. The Goze-decomposition is interesting from a theoretical point of view: it is a natural development, to be associated with every infinitesimal vector, without any preliminary condition.

The Goze decomposition has farreaching applications in notably algebraic geometry and in the theory of Lie-Algebra's, see [27] [28].
Here we will only point out that the Goze-decomposition yields shadow expansions (Section 3).

2) Of course the vectors V_1 and V_2 are not unique. But the carrier of V_1 is determined without ambiguity, for it is the shadow of the carrier of $\binom{\varepsilon_1}{\eta_1}$. The vector $\varepsilon_2 V_1$ is a first approximation of $\binom{\varepsilon_1}{\eta_1}$, indeed

$$\left|\binom{\varepsilon_1}{\eta_1}-(\varepsilon_2 V_1)\right| = \emptyset.\left|\binom{\varepsilon_1}{\eta_1}\right|$$

See Fig. 1.

3) The Goze decomposition of an infinitesimal vector can be seen as an extension of the notion of Taylor expansion. Suppose $\eta_1 = f(\varepsilon_1)$, where f is a standard C_1 function. Then $\eta_1 = \varepsilon_1 f'(0)+\eta_2\varepsilon_1$ with $\eta_2 \simeq 0$. So one has the Goze decomposition

$$\binom{\varepsilon_1}{\eta_1} = \varepsilon_1\binom{1}{f'(0)} + \varepsilon_1\eta_2\binom{0}{1}$$

As such, it is a strict extension. The function $f(x) = \sqrt{x}$ does not possess a Taylor development (of order 1) in 0. However,

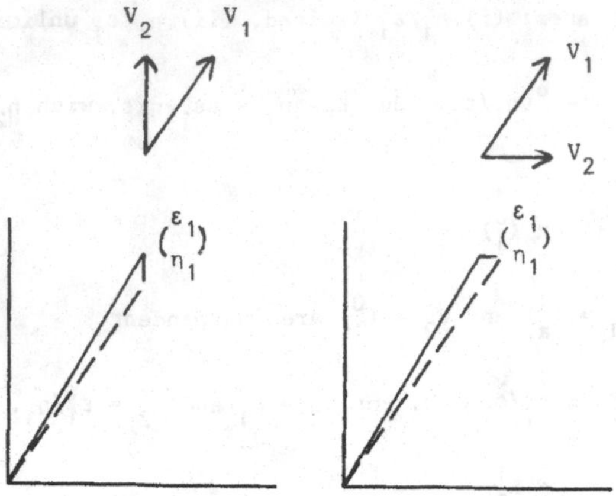

Figure 1: If η_1/ε_1 is appreciable one has two major choices in V_1 and V_2, i.e. one may have

$$\binom{\varepsilon_1}{\eta_1} = \varepsilon_1\binom{1}{a} + \eta_2\varepsilon_1\binom{0}{1} \qquad (a = {}^{o}(\eta_1/\varepsilon_1);\ \text{left})$$

$$\binom{\varepsilon_1}{\eta_1} = \eta_1\binom{1/a}{1} + \eta_2\eta_1\binom{1}{0} \qquad (a = {}^{o}(\eta_1/\varepsilon_1);\ \text{right})$$

The first vectors have always the same carrier.

$$\binom{\varepsilon_1}{\sqrt{\varepsilon_1}} = \sqrt{\varepsilon_1}\binom{0}{1} + \sqrt{\varepsilon_1}\cdot\sqrt{\varepsilon_1}\binom{1}{0}$$

4) The vectors $\binom{\varepsilon_1}{\eta_1}$ and $\binom{\varepsilon_2}{\eta_2}$ are related by a standard quadratic transformation $T: \mathbb{R}^2 \to \mathbb{R}^2$, namely $T\binom{X}{Y} = (V_1\ V_2X)\binom{X}{Y}$. Because there is some freedom in the choice of the vector $\binom{\varepsilon_2}{\eta_2}$, this transformation is not unique. But it follows from the above proof that T can always be choosen to be of one of the following forms:

1)	$T(X,Y) = (X, aX+XY)$	if η_1/ε_1 is limited
2)	$T(X,Y) = (XY, X)$	if η_1/ε_1 is unlimited.

5) The Goze decomposition can be continued: the vector $\binom{\varepsilon_2}{\eta_2}$ may be decomposed to give $\binom{\varepsilon_2}{\eta_2} = \varepsilon_3V_3+\varepsilon_3\eta_3V_4$, the vector $\binom{\varepsilon_3}{\eta_3}$ will given some $\binom{\varepsilon_4}{\eta_4}$.... . So, by the foregoing remark, and by standardization, it is seen that to each infinitesimal vector $\binom{\varepsilon}{\eta}$ is associated a standard sequence of quadratic transformations $(T_n)_{n\geq 1}$, each T_n being of the form 1) or 2).

In particular, if $\eta = f(\varepsilon)$, where f is a standard function of class C^n(st n), then every transformation can be taken of the form 1), and one obtains

$$\binom{\varepsilon}{\eta} = \begin{pmatrix}1 & 0\\ f(0) & \varepsilon\end{pmatrix}\begin{pmatrix}1 & 0\\ \frac{f'(0)}{2} & \varepsilon\end{pmatrix}\cdots\begin{pmatrix}1 & 0\\ \frac{f^{(n)}(0)}{n!} & \varepsilon\end{pmatrix}\binom{\varepsilon}{\eta_n} \quad \text{with } \eta_n \simeq 0$$

6) An "algorithmic" way to get a particular kind of Goze decompositions is the following. Given an infinitesimal vector $\binom{\varepsilon_1}{\eta_1}$, the component ε_2 of the new infinitesimal vector $\binom{\varepsilon_2}{\eta_2}$ is defined by $\varepsilon_2 = \varepsilon_1$ if $|\varepsilon_1| \geq |\eta_1|$ and $\varepsilon_2 = \eta_1$ if $|\eta_1| > |\varepsilon_1^2|$. Then there are two possibilities: $V_1 = \binom{1}{a}$, $V_2 = \binom{0}{1}$ with $|a| = |{}^o(\eta_1/\varepsilon_1)| \leqq 1$ or $V_1 = \binom{b}{1}$, $V_2 = \binom{1}{0}$ with $|b| = |{}^o(\eta_1/\varepsilon_1)| \leqq 1$.

7) The Goze decomposition can be extended to infinitesimal vectors $(\alpha_1,\ldots,\alpha_n)^T$ of $\mathbb{R}^n$, for st n. Indeed, using the procedure described in 6), it is easily seen that there exist n linear independent standard vectors $V_1,\ldots,V_n$ and n infinitesimals $\beta_1,\ldots,\beta_n$ such that

$$\begin{pmatrix}\alpha_1\\ \vdots\\ \alpha_n\end{pmatrix} = \beta_1 V_1 + \beta_1\beta_2 V_2 + \ldots + \beta_1\beta_2\ldots\beta_n V_n$$

The Goze decomposition may even be generalized to limited vectors M. Indeed, first take the shadow ${}^oM$, and then apply the Goze decomposition to the infinitesimal vector $M-{}^oM$.

2. Scales.

Definition 6.2(i): An internal sequence of real numbers $(u_n)_{n\leqq k}$, $k \in \mathbb{N}$, will be called a finite order scale if $u_{n+1}/u_n \simeq 0$ for all $n < k$.

(ii) An internal sequence of real numbers $(u_n)_{n\in\mathbb{N}}$ will be called an order scale if $u_{n+1}/u_n \simeq 0$ for all $n \in \mathbb{N}$.

Examples. (1) Let ω be positive unlimited. Then $(\omega, \log\omega, 1)$ is a finite order scale.

(2) Let $\varepsilon \simeq 0$, $\varepsilon \neq 0$. Then the sequence $(\varepsilon^n)_{n\in\mathbb{N}}$ is an order scale.

3) Let $\varepsilon \simeq 0$, $\varepsilon > 0$. Put $u_n = \varepsilon^{-1/n}$. Then $u_{n+1}/u_n \simeq 0$ for all st $n \geqq 1$. By Robinson's lemma, there exists some unlimited $\omega \in \mathbb{N}$ such that $u_{n+1}/u_n \simeq 0$ up to ω. Then $(u_1,\ldots,u_\omega)$ is a finite order scale.

Which numbers should figure in an order scale? One would like to compare the order of magnitude of some real number with a real number whose order of magnitude is "known". As far as real numbers of the form $f(\omega)$ are concerned, where $f: \mathbb{R} \to \mathbb{R}$ is standard and ω is unlimited, a partial answer to the forewarded question seems to be possible on account of observations made by G.H. Hardy. In [46] he introduces the notion of L-functions. A L-function is a real function, defined on some interval $[a,\infty[$ by a finite combination of the symbols +, −, ., ___ (division), $\wedge$ (exponentiation), log and exp, and real constants. For instance

$$x^{1/2},\ x \log (x^2-5),\ e^{e^{x+1}}$$

are L-functions. Hardy shows that every two L-functions φ and Ψ may be compared in the following sense: $\lim_{x \to +\infty} \varphi(x)/\Psi(x)$ exists or $\lim_{x \to +\infty} \Psi(x)/\varphi(x)$ exists.

Now consider $F = \{st\ f \mid f: \mathbb{R}^+ \to \mathbb{R}\}$. Let us define an equivalence relation on this external set by $f(\omega) \sim g(\omega) \Leftrightarrow f(\omega)/g(\omega) \in \not{A}$. Consider a choice function on these equivalence classes; for each class a number $\overline{f}(\omega)$ is thus defined. One could ask $\overline{f}$ to be a L-function -if available- which should be of the "simplest" form. For instance, one could prefer $\log \omega$ above $\log(\omega+1)$ (Though in some cases $\log(\omega+1)$ could be prefered). Then by Hardy's theorem, for each standard L-function φ there exists some function $\overline{f}$ such that ${}^o(\varphi(\omega)/\overline{f}(\omega))$ exists (and is different from 0, and does not depend on ω).

If there are no L-functions available in some class, one could prefer numbers $d(\omega)$, where d is some standard <u>order function</u>, i.e. a monotonous positive real function.

But such numbers are not present in all classes. Take for example $a_\omega = e^\omega \sin \omega$. If $\omega = 2\pi\nu$, $\nu \in \mathbb{N}$ then $a_\omega = 0$; if $\omega = 2\pi(\nu+1/4)$ then $a_\omega = e^\omega$. Now put $S_\omega = \bigcap_{f\in F, f \text{ increasing}} [0,f(\omega)]-\not{\mathbb{C}}^+$. By Theorem 4.17(ii), S_ω is not empty. Furthermore, $S_\omega = S_{\omega'}$, for all $\omega' \in [2\pi\nu, 2\pi(\nu+1/4)]$. By continuity there is $\overline{\omega}$ such that $a_{\overline{\omega}} \in S_\omega$. Hence a_ω cannot be compared with any number $d(\omega)$, where d is a standard order function.

From now on we will always assume that an order scale is constituted of <u>positive</u> real numbers.

We proceed to show that there are two kinds of scales.

<u>Definition 6.3</u>: Let $(u_n)_{n\in\mathbb{N}}$ be an order scale. The set

$$M = \bigcap_{n\in\mathbb{N}} [-u_n, u_n]$$

will be called the <u>microhalo</u> associated to the scale $(u_n)_{n\in\mathbb{N}}$. Sometimes we write u-M.

By proposition 4.31 every microhalo is a nonlinear halo (and vice-versa). We already came across the microhalo associated to the scale $(\varepsilon^n)_{n\in\mathbb{N}}$: the <u>ε-microhalo</u> $\varepsilon\text{-M} = \bigcap_{n\in\mathbb{N}} [-\varepsilon^n,\varepsilon^n]$.

<u>Definition 6.4</u>: Let M be the microhalo associated to an order scale $(u_n)_{n\in\mathbb{N}}$. Put

$$L = \{x \mid x\mu \in M \text{ for all } \mu \in M\}$$

An element x of L will be called <u>asymptotically limited</u> with respect to the scale $(u_n)_{n\in\mathbb{N}}$.

<u>Comments</u>: 1) The notion "asymptotically limited" is inspired by Van der Corput's definition of "asymptotically finite" in [41]. He uses it to calculate some limits. Notice that the use of the term "limited" agrees with the position of the limited numbers as regards to the linear halos.

2) Here are some examples. The number ω is always positive unlimited.

M	L
$\bigcap_{n\in\mathbb{N}} [-1/\omega^n, 1/\omega^n]$	$\bigcup_{n\in\mathbb{N}} [-\omega^n,\omega^n]$
$\bigcap_{n\in\mathbb{N}} [-\frac{\omega}{\log^n\omega}, \frac{\omega}{\log^n \omega}]$	$\bigcup_{n\in\mathbb{N}} [-\log^n \omega, \log^n \omega]$
$\bigcap_{n\in\mathbb{N}} [-\omega^{1/n}, \omega^{1/n}]$	$\bigcap_{n\in\mathbb{N}} [-\omega^{1/n}, \omega^{1/n}]$
$\bigcap_{n\in\mathbb{N}} [-\log_n \omega, \log_n \omega]$	$\bigcap_{n\in\mathbb{N}} [-\log_n \omega, \log_n \omega]$

3) The following facts are clearly verified. The set of asymptotically limited numbers with respect to some scale is a convex ring with unity, and contains $\mathbf{\mathfrak{L}}$ strictly (4.31). Homothetical scales $(u_n)_{n\in\mathbb{N}}$ and $(ru_n)_{n\in\mathbb{N}}$, where $r > 0$, have the same set of asymptotically limited numbers.

The above examples suggest the following proposition.

<u>Proposition 6.5</u>: The set of all asymptotically limited numbers with respect to some order scale is either a nonlinear galaxy, or a nonlinear halo.

<u>Proof</u>: Let M be the microhalo associated to the scale, and L be its set of asymptotically limited numbers. Put $H = \log(M^+-\{0\})$. As a consequence of the theorem of classification of cuts (4.34) either there exists $a > H$ and a group G which is a galaxy, such that $H =]-\infty,a]\setminus a+G$, or there exists $b \in H$, and a group K which is a halo, such that $H =]-\infty,b] \cup b+K$. One shall see that in the first case

$L^+\setminus[0,1[\ = \exp(G^+)$ -and L is a galaxy-, and in the second case $L^+\setminus[0,1[\ = \exp(K^+)$ -and L is a halo-. We treat only the first case, the proof of the second case being similar.

So let $g \in \exp(G^+)$ and $\mu \in M^+-\{0\}$. We must show that $g\mu \in M$. Now $\log(g\mu) = \log\mu + \log g$. Because $\log\mu \in H$ and $\log g \in G$, which acts as thickness of the border for the halo H, one has $\log\mu + \log g \in H$. Hence $g\mu \in M$. Hence $\exp(G^+) \subset L^+\setminus[0,1[$.

Conversely, let $l \in L^+-[0,1[$. We must show that log l belongs to the thickness of the border of the halo H. Indeed, let $h \in H$. Then $e^h \in M$, so $le^h \in M$. Hence $h + \log l \in H$. Because $\log l \geq 0$, the number log l is element of the thickness of the border of H, and thus of G^+. So $l \in \exp(G)$. Hence $L^+-[0,1[\ \subset \exp(G^+)$.

Combining, we obtain $L^+-[0,1[\ = \exp(G^+)$. Hence L is a galaxy. By comment 3) L contains some unlimited number ω, and L is a ring. Hence $\omega L \subset L$. By proposition 4.29 the galaxy L is nonlinear.

So there exist two sorts of order scales: scales whose set of asymptotically limited numbers is a nonlinear galaxy, and scales whose set of asymptotically limited numbers is a nonlinear halo.

3. Definition and fundamental properties of shadow expansions.

Definition and examples.

Definition 6.6: Let $a \in \mathbb{R}$ and $(u_k)_{k\in\mathbb{N}}$ be an order scale. Let $n \in \underline{\mathbb{N}}$. We say that a possesses a shadow expansion of order n with respect to this scale if there exist standard numbers $c_0,\dots,c_n$ such that

$$\frac{a-\sum_{k=0}^{n-1} c_k u_k}{u_n} \simeq c_n$$

Then we may write $a = \sum_{k=0}^{n} c_k u_k + \emptyset . u_n$.

Comment. If a possesses a shadow expansion of order n with respect to a scale $(u_n)_{n\in\mathbb{N}}$, then a possesses a shadow expansion of order m, with respect to this scale, for all m < n. Indeed

$$\frac{a-\sum_{k=0}^{m-1} c_k u_k}{u_m} = c_m + c_{m+1}\frac{u_{m+1}}{u_m} + \dots + c_n\frac{u_n}{u_m} \simeq c_m$$

In fact, shadow expansions are defined by external induction: $c_0 = {}^o(a/u_0)$, and if $c_0,\dots,c_{n-1}$ are defined, then $c_n = \dfrac{{}^o(a-\sum_{k=0}^{n-1} c_k u_k)}{u_n}$. So the expansion is obtained by taking successive shadows.

Because the shadow of a number, if it exists, is unique, the constants $c_0,\dots,c_n$ are unique.

Definition 6.7: We say that a number a possesses a shadow expansion with respect to an order scale if a possesses a shadow expansion of order n with respect to this scale for all $n \in \mathbb{N}$.

Notation: The shadow expansion of a number a with respect to a scale $(u_n)_{n\in\mathbb{N}}$ determines a sequence of standard constants $(c_n)_{n\in\underline{\mathbb{N}}}$. By standardization, this sequence can be continuated into a standard sequence $(c_n)_{n\in\mathbb{N}}$. This being a purely formal operation, the sum $\sum_{n=0}^{\infty} c_n u_n$ does not need to converge. If $\sum_{n=0}^{\infty} c_n u_n$ converges to a we write

$$a = \sum_{n=0}^{\infty} c_n u_n$$

If $\sum_{n=0}^{\infty} c_n u_n$ does not converge to a, diverges, or if we wish to leave this undecided, then we write

$$a \sim \sum_{n=0} c_n u_n$$

Definition 6.8: Shadow expansions with respect to a scale $(\varepsilon^n)_{n\in\mathbb{N}}$, $\varepsilon \simeq 0$ are called expansions in ε-shadow (after F. Diener [18]).

Examples: The number ε will always be considered strictly positive and infinitesimal.

1) The number $\frac{1}{1+\varepsilon}$ possesses the expansion in ε-shadow $\sum_{n=0}^{\infty} (-1)^n\varepsilon^n$. Indeed, for all st n,

$$\frac{\frac{1}{1+\varepsilon} - \sum_{k=0}^{n-1} (-1)^k\varepsilon^k}{\varepsilon^n} = \frac{(-1)^n}{1+\varepsilon} \simeq (-1)^n$$

Furthermore the sum $\sum_{n=0}^{\infty} (-1)^n\varepsilon^n$ converges to the number $\frac{1}{1+\varepsilon}$.

In general, if a standard function f(x) possesses a Taylor expansion $\sum_{n=0} a_n x^n$, then f(ε) possesses the expansion in ε-shadow $f(\varepsilon) \sim \sum_{n=0} a_n\varepsilon^n$ (see Proposition 2.6).

2) The number $e^{-1/\varepsilon}$ possesses the expansion in ε-shadow $\sum_{n=0} 0$, for $e^{-1/\varepsilon}/\varepsilon^n \simeq 0$ for all st n. This example shows that two different numbers may have the same shadow expansion, for also $0 \sim \sum_{n=0} 0$.

3) The number $\frac{1}{1+\varepsilon+\varepsilon^{3/2}}$ possesses an expansion in ε shadow of order 1, but not of order 2, for

$$\frac{1}{1+\varepsilon+\varepsilon^{3/2}} = 1-\varepsilon-\varepsilon^{3/2}(1+\emptyset)$$

However, it possesses a full expansion in $\varepsilon^{1/2}$-shadow.

4) Let f be a standard real C^{∞} function on $[0,\infty[$ of exponential order. Let ω be positive and unlimited. Let F(x) be the Laplace transform

$$F(x) = {}_0\int^{\infty} e^{-xt} f(t)dt$$

Then $F(\omega)$ possesses the expansion in $1/\omega$-shadow

$$F(\omega) \sim \sum_{n=0} f^{(n)}(0)/\omega^{n+1}$$

This follows from Watson's lemma (2.9).

5) Consider the finite order scale $(1/\varepsilon,1)$, and the Riemann ζ-function $\zeta(s) = \sum_{n=1}^{\infty} \frac{1}{n^s}$. We will show that the number $\zeta(1+\varepsilon)$ possesses the shadow expansion

$$\zeta(1+\varepsilon) = 1/\varepsilon+\Gamma+\alpha$$

where Γ is the Euler constant and $\alpha \simeq 0$. Indeed, for each st k $\sum_{n=1}^{k} \frac{1}{n^{1+\varepsilon}} \simeq \sum_{n=1}^{k} \frac{1}{n}$. By Robinson's lemma, there exists some unlimited $\omega \in \mathbb{N}$ such that $\sum_{n=1}^{\omega} \frac{1}{n^{1+\varepsilon}} \simeq \sum_{n=1}^{\omega} \frac{1}{n}$. We may suppose that $\varepsilon \log^2\omega \simeq 0$. Then, using Lemma 5.9

$$\sum_{n=1}^{\infty} \frac{1}{n^{1+\varepsilon}} = \sum_{n=1}^{\omega-1} \frac{1}{n^{1+\varepsilon}} + \sum_{n=\omega}^{\infty} \frac{1}{n^{1+\varepsilon}}$$

$$\simeq \sum_{n=1}^{\omega-1} \frac{1}{n} + {}_{\omega}\int^{\infty} \frac{1}{t^{1+\varepsilon}}dt$$

$$= \log\omega + \Gamma + \frac{\omega^{-\varepsilon}}{\varepsilon}$$

$$= \log\omega + \Gamma + \frac{e^{-\varepsilon\ \log\omega}}{\varepsilon}$$

$$= \log\omega + \Gamma + \frac{(1-\varepsilon\log\omega + \frac{\varepsilon^2\log^2\omega(1+\emptyset)}{2})}{\varepsilon}$$

$$= 1/\varepsilon + \Gamma + \emptyset$$

(This argument is due to J.-L. Callot, of course the relation $\zeta(1+x) = 1/x+\Gamma+o(1)$ for $x \to 0$ is classical).

6) The logarithmic integral li(x) is defined by

$$\mathrm{li}(x) = {}_0\!\!\int^{x} \frac{1}{\log t}dt$$

For positive unlimited ω, the number $\mathrm{li}(\omega)$ has the shadow expansion

$$\mathrm{li}(\omega) \sim \sum_{n=1} (n-1)!\frac{\omega}{(\log \omega)^n}$$

This follows from Proposition 3.3, by means of the identity $\mathrm{li}(x) = \frac{x}{\log x}\ {}_0\!\!\int^{x} \frac{e^{-u}}{1-u/\log x}du$. Here we give a direct proof. Firstly, let us write

$$\mathrm{li}(\omega) = {}_0\!\!\int^{a} \frac{1}{\log t}dt + {}_a\!\!\int^{\omega} \frac{1}{\log t}dt \qquad \text{(st } a > 1\text{)}$$

Secondly, by integration by parts

$${}_a\!\!\int^{\omega} \frac{1}{\log t}dt = \sum_{k=1}^{n-1} \omega\frac{(k-1)!}{(\log \omega)^k} - \sum_{k=1}^{n-1} a\frac{(k-1)!}{(\log a)^k} + (n-1)!\ {}_a\!\!\int^{\omega} \frac{1}{(\log t)^n}dt$$

We first show that ${}_a\!\int^{\omega} \frac{1}{(\log t)^n}dt = (1+\emptyset)\frac{\omega}{(\log \omega)^n}$. Indeed, let $\nu > 0$ be unlimited such that $\nu < \frac{\omega}{(\log \omega)^n}$ for all $n \in \mathbb{N}$. Now

$${}_a\!\!\int^{\omega} \frac{1}{(\log t)^n}dt = \frac{t}{(\log t)^n}\Big|_a^{\omega} + n\ {}_a\!\!\int^{\omega} \frac{1}{(\log t)^{n+1}}dt$$

$$= \frac{\omega}{(\log \omega)^n} - \frac{a}{(\log a)^n} + n\ {}_a\!\!\int^{\nu} \frac{dt}{(\log t)^{n+1}} + n\ {}_\nu\!\!\int^{\omega} \frac{dt}{(\log t)^{n+1}}$$

Now $\frac{a}{(\log a)^n} = \emptyset.\frac{\omega}{(\log \omega)^n}$; $n\ {}_a\!\int^{\nu} \frac{dt}{(\log t)^{n+1}} \leq n\frac{\nu}{(\log a)^{n+1}} = \emptyset.\frac{\omega}{(\log \omega)^n}$, and

$$n\ {}_\nu\!\!\int^{\omega} \frac{dt}{(\log t)^{n+1}} \leq \frac{n}{\log \nu}\ {}_\nu\!\!\int^{\omega} \frac{dt}{(\log t)^n} = \emptyset.\ {}_a\!\!\int^{\omega} \frac{dt}{(\log t)^n}.$$ Hence

${}_a\!\int^{\omega} \frac{1}{(\log t)^n}dt = (1+\emptyset)\frac{\omega}{(\log \omega)^n}$. Also, $\sum_{k=1}^{n-1} \frac{a(k-1)!}{(\log a)^k} = \emptyset.\frac{\omega}{(\log \omega)^n}$, and ${}_0\!\int^{a} \frac{1}{\log t}dt =$ $\emptyset.\frac{\omega}{(\log \omega)^n}$. So we find

$$\mathrm{li}(\omega) = \sum_{k=1}^{n} (k-1)!\frac{\omega}{(\log \omega)^k} + \emptyset.\frac{\omega}{(\log \omega)^n}$$

Existence theorems.

Proposition 6.9: Let $(u_n)_{n\in\mathbb{N}}$ be an order scale and M be its microhalo. Then

$$M = \{a \mid a \sim \sum_{n=0} 0\}$$

Proof: Let $a \in M$ and $n \in \mathbb{N}$. Then $|a| \leq u_{n+1}$, so $a/u_n \simeq 0$. Hence $a \sim \sum_{n=0} 0$.
Conversely, let $a \sim \sum_{n=0} 0$.

Then $a/u_n = \dfrac{a - \sum_{k=0}^{n-1} 0}{u_n} \simeq 0$ for all st n. Hence $a \in M$.

Corollary 6.10: Let $(u_n)_{n\in\mathbb{N}}$ be an order scale. Suppose $a \sim \sum_{n=0} c_n u_n$ for some standard sequence $(c_n)_{n\in\mathbb{N}}$. Then

$$a+M = \{x \mid x \sim \sum_{n=0} c_n u_n\}$$

So a number having a shadow expansion with respect to a scale is determined up to an element of the microhalo of this scale. Given a formal series $\sum_{n=0} c_n u_n$, does it represent the shadow expansion of a real number? Next theorem states that the answer is positive, and even more, stands very close to the formal series, for the number may be one of its partial sums.

Theorem 6.11: (existence theorem). Let $(u_n)_{n\in\mathbb{N}}$ be an order scale and $(c_n)_{n\in\mathbb{N}}$ be a standard sequence. Then there exists an unlimited index ω_c such that for all unlimited $\omega \leq \omega_c$

$$\sum_{n=0}^{\omega} c_n u_n \sim \sum_{n=0} c_n u_n$$

Proof: As long as n is standard, the quotient of two successive nonzero terms of the sequence $\sum_{n=0} c_n u_n$ is infinitesimal. By Robinson's lemma this property is preserved up to some unlimited index ω_c. Let $\omega \leq \omega_c$ be unlimited. Then, using Lemma 5.12, one has for all st $n \in \mathbb{N}$

$$\frac{\sum_{n=0}^{\omega} c_k u_k - \sum_{k=0}^{n-1} c_k u_k}{u_n} = \sum_{k=n}^{\omega} c_k u_n/u_k \simeq c_n$$

Hence $\sum_{n=0}^{\omega} c_n u_n \sim \sum_{n=0} c_n u_n$.

Comment: The proof can be made more constructive. For instance, let ω_c' be the largest index such that all quotients of two successive non-zero terms of lesser indices is smaller than 1/2 (if defined, otherwise take $\omega_s = \omega$). Again using 5.12, one obtains $\sum_{k=0}^{\omega} c_k u_k \sim \sum_{k=0} c_k u_k$ for all $\omega \leqq \omega_c'$.

The following refinement has been suggested by G. Reeb (private communication).

Theorem 6.12. Let $(u_n)_{n\in\mathbb{N}}$ be an order scale. Then there exists an unlimited index $\rho \in \mathbb{N}$ such that

$$\sum_{n=0}^{\rho} c_n u_n \sim \sum_{n=0} c_n u_n$$

for all standard sequences $(c_n)_{n\in\mathbb{N}}$. In fact, $\sum_{n=0}^{\omega} c_n u_n \sim \sum_{n=0} c_n u_n$ for all unlimited indices ω up to ρ.

Proof: By the existence theorem, for every standard sequence $(c_n)_{n\in\mathbb{N}}$ there exists an unlimited number $\omega_c \in \mathbb{N}$ such that $\sum_{n=0}^{\omega} c_n u_n \sim \sum_{n=0} c_n u_n$ for all unlimited $\omega \leqq \omega_c$. By the principle of the universal index (4.20) there exists $\rho \in \mathbb{N}$ which is less than ω_c for all standard c, but is still unlimited. Then $\sum_{n=0}^{\omega} c_n u_n \sim \sum_{n=0} c_n u_n$ for all unlimited $\omega \leqq \rho$, for all standard sequences c.

Put $S_\rho = \sum_{n=0}^{\rho} c_n u_n$. It is tempting to consider S_ρ as a "sum" to the formal series $\sum_{n=0} c_n u_n$, the index ρ being valid for all standard sequences.

Let us combine the results 6.10–6.12 into the following fundamental theorem.

Theorem 6.13: Let $(u_n)_{n\in\mathbb{N}}$ be an order scale. Then there exists an unlimited index ρ such that for all standard sequences $(c_n)_{n\in\mathbb{N}}$

$$\sum_{n=0}^{\rho} c_n u_n + u\text{-}M = \{x \mid x \sim \sum_{n=0} c_n u_n\}$$

So all numbers a having a shadow expansion with respect to some scale $(u_n)_{n\in\mathbb{N}}$ have a representation

$$a = S_\rho + \mu \qquad (\rho \in \mathbb{N} \text{ unlimited}, \mu \in u\text{-}M)$$

Comment: The existence theorem had already been proved by F. Diener. We presented a different proof here. Concerning classical asymptotic expansions, existence proofs have been given by among others. Borel, Ritt [47] and Van der Corput [40].

Shadow expansions of functions.

We will now briefly consider shadow expansions of functions. First we recall the definition of the shadow of a real function. A real function f is said to be of class S_0 on an internal subset $E \subset \mathbb{R}$ if f takes limited values for all limited elements of E and if $f(x) \simeq f(y)$ for all limited $x,y \in E$ such that $x \simeq y$. Let f: $E \to \mathbb{R}$ be a function of class S_0 on E. Then one may define for all st $x \in E$

$$g(x) = {}^{o}(f(x))$$

By standardization g can be extended to a function f: ${}^{S}E \to \mathbb{R}$. The function g is called the <u>shadow</u> of the function f. It can be shown ([1], [3]) that $g(x) \simeq f(x)$ for all limited $x \in {}^{S}E \cap E$ and that g is continuous on ${}^{S}E \cap E$. Conversely, if the shadow -i.e. the standardized of the halo- of the graph of a real function f which takes limited values for limited arguments, is the graph of a standard <u>function</u>, the function f is of class S_o. See also Fig. 2.

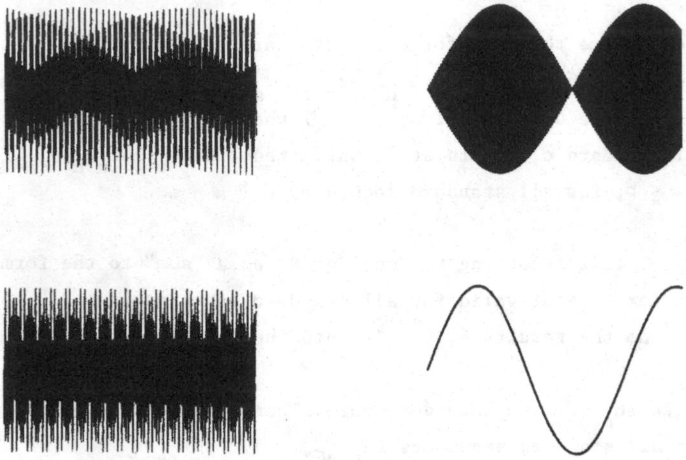

<u>Figure 2</u>: Only functions f of class S_0 possess a shadow, i.e. a standard function ${}^{o}f$ which is infinitely close to it for limited arguments. The standard function ${}^{o}f$ is defined by taking into consideration the values of f for only the standard reals x: $({}^{o}f)(x) = {}^{o}(f(x))$. Its domain is extended to the whole $\mathbb{R}$ by the axiom of standardization.

The function sin ωx is not of class S_0 for unlimited ω, for it is not S-continuous. So the standard function g defined by $g(x) = {}^{o}\sin \omega x$, for st $x \in \mathbb{R}$, may very well not look alike the original function.

In the above drawing we did something which bears some resemblance to the operation of "taking the shadow". We drew sin ωx for $x = \frac{n}{\omega}$, $0 \leq n \leq 300$, and ω = 270,316,384,631, respectively.

For reasons of simplicity, we take from now on $E = \mathbb{R}$.

<u>Definition 6.14</u>: Let $(u_k)_{k\in\mathbb{N}}$ be an order scale, let f be a real internal function and let st $n \in \mathbb{N}$. We say that f possesses a <u>shadow expansion</u> of order n on $\mathbb{R}$ with respect to the scale $(u_k)_{k\in\mathbb{N}}$ if there exist standard real functions $f_0,\dots,f_n$ such that

$$^o\left(\frac{f(x)-\sum_{k=0}^{n-1} f_k(x)u_k}{u_n}\right) = f_n(x)$$

for all standard $x \in \mathbb{R}$.

A function f is said to possess a <u>shadow expansion</u> on $\mathbb{R}$ with respect to the scale $(u_k)_{k\in\mathbb{N}}$ if f possesses a shadow expansion with respect to this scale for all st $n \in \mathbb{N}$. We then write $f \sim \sum_{n=0} f_n u_n$.

If $(u_k)_{k\in\mathbb{N}} = (\varepsilon^k)_{k\in\mathbb{N}}$, where $\varepsilon > 0$, $\varepsilon \simeq 0$, we speak of an <u>expansion in ε-shadow</u>, after F. Diener.

<u>Comment</u>: If $^o(f(x)-\sum_{k=0}^{n-1} f_k(x)u_k)/u_n$ exists for all st $x \in \mathbb{R}$, then $f_n(x)$ is continuous on $\mathbb{R}$, and $(f(x)-\sum_{k=0}^{n-1} f_k(x)u_k)/u_n$ is of class S_0 on $\mathbb{R}$. Hence, for all limited $y \in \mathbb{R}$, with $x = {}^oy$:

$$\frac{f(y)-\sum_{k=0}^{n-1} f_k(y)u_k}{u_n} \simeq \frac{f(x)-\sum_{k=0}^{n-1} f_k(x)u_k}{u_n} \simeq f_n(x) \simeq f_n(y)$$

Thus

$$f(y) = \sum_{k=0}^{n} f_k(y)u_k + \emptyset.u_n$$

<u>Examples</u>: All examples concern expansions in ε-shadow expansion

1) The function $f(x) = \frac{1}{1+\varepsilon x}$ possesses the shadow

$$\frac{1}{1+\varepsilon x} = \sum_{n=0}^{\infty} (-x)^n \varepsilon^n$$

In general, if f is standard C^∞, then $f(\varepsilon x)$ possesses the shadow expansion

$$f(\varepsilon x) = \sum_{n=0} \frac{f^{(n)}(0)}{n!} x^n.\varepsilon^n$$

2) Suppose a function f has an expansion in ε-shadow on some standard interval [a,b]. Put

$$F = \int_a^b f(t)dt$$

Then the number F has an expansion in ε-shadow. Cf. Watson's lemma.

3) Expansions in ε-shadow of functions are of particular interest in the nonstandard approach of the study of singular perturbations. For instance, if in the differential equation

$$\varepsilon\frac{dy}{dx} = f(x,y)$$

the function f has an expansion in ε-shadow, then under fairly general conditions a solution $\overline{y}(x)$ has an expansion in ε-shadow. See [18], [20].

We now prove an existence theorem.

<u>Theorem 6.15</u>: Let $(u_n)_{n\in\mathbb{N}}$ be an order scale and $(f_n)_{n\in\mathbb{N}}$ be a standard sequence of continuous real functions, defined on $\mathbb{R}$. Then there exists a real function f such that

$$f \sim \sum_{n=0} f_n u_n$$

<u>Proof</u>: Let ρ be a universal index for the scale $(u_n)_{n\in\mathbb{N}}$ (Theorem 6.12). Then $\sum_{n=0}^{\rho} f_n(x)u_n \sim \sum_{n=0} f_n(x)u_n$ for all standard real numbers x. Hence define $f: \mathbb{R} \to \mathbb{R}$ by $f(y) = \sum_{n=0}^{\rho} f_n(y)u_n$. Then $f \sim \sum_{n=0} f_n u_n$.

<u>Comment</u>: Let φ be a real function having a shadow expansion $\varphi(x) \sim \sum_{n=0} f_n(x)u_n$ where $(u_n)_{n\in\mathbb{N}}$ is an order scale and $(f_n)_{n\in\mathbb{N}}$ is a standard sequence of (continuous) functions. We will show that there is a representation

$$\varphi(x) = \sum_{k=0}^{\rho} f_k(x)u_k + \mu(x)$$

where ρ is a universal index, common to all shadow expansions of functions with respect to the scale $(u_n)_{n\in\mathbb{N}}$, and μ is an internal function such that $\mu(x) \in u\text{-}M$ for all limited x. (Theorem 6.15 only implies that $\mu(x) \in u\text{-}M$ for st x). Indeed, let K be any positive unlimited number such that $u_{n+1}/u_n \leqq 1/K$ for all st n, and ν be an unlimited index such that $u_{n+1}/u_n \leqq 1/K$ for all $n \leqq \nu$; for every standard sequence $f = (f_n)_{n\in\mathbb{N}}$ there exists some unlimited index ν_f such that $f_n(x)/K \simeq 0$ for all limited x (all these affirmations follow from permanence principles); By the principle of the universal index, there exists some unlimited index ρ such that $\rho \leqq \nu$ and $\rho \leqq \nu_f$ for all standard sequences f.

For all st n the internal function $\alpha_n(x) = (\varphi(x) - \sum_{k=0}^{n} f_k(x)u_k)/u_n$ has the property that $\alpha_n(x) \simeq 0$ for all limited x. By the multidimensional permanence principle (4.19) the property "$\alpha_n(x) \simeq 0$ for all limited x" still holds up to some unlimited ω (assumed $\leqq \rho$). Put $\mu(x) = \varphi(x) - \sum_{k=0}^{\rho} f_k(x)u_k$. Then for all limited x

$$\mu(x) = \varphi(x) - \sum_{k=0}^{\omega} f_k(x)u_k + u_{\omega+1} \sum_{k=\omega+1}^{\rho} f_k(x)\frac{u_k}{u_{\omega+1}}$$

$$= \emptyset.u_\omega + \emptyset.K.u_{\omega+1} \in u\text{-}M$$

One may not ask that $\mu(x) \in u\text{-}M$ for all $x \in \mathbb{R}$: the function $f(x) = e^{-\frac{1}{\varepsilon|x|}}$ ($\varepsilon > 0$, $\varepsilon \simeq 0$) possesses an expansion in ε-shadow, namely $f(x) \sim \sum_{n=0} 0$, but $f(1/\varepsilon) \notin \varepsilon\text{-}M$.

Goze decomposition and shadow expansions.

The following proposition establishes a relation between the Goze-decomposition and the expansion in ε-shadow.

Proposition 6.16: Let $\varepsilon,\eta \simeq 0$, st $n \in \mathbb{N}$. The number η possesses an expansion in ε-shadow if and only if there are infinitesimals $\eta_0 = \eta, \eta_1, \ldots, \eta_n$ such that $\binom{\varepsilon}{\eta_{k-1}} = T_k\binom{\varepsilon}{\eta_k}$ where $T_k: \mathbb{R}^2 \to \mathbb{R}^2$ is a standard quadratic transformation of the form $(X,Y) \to (X,(a_k+Y)X)$ for $1 \leq k \leq n$.

Proof: First assume that $\eta = a_1\varepsilon + \ldots + a_n\varepsilon^n + \eta_n\varepsilon^n$, st $a_1,\ldots,$st a_n, $\eta_n \simeq 0$. Define for $0 \leq k \leq n-1$

$$\eta_k = a_{k+1}\varepsilon + a_{k+2}\varepsilon^2 + \ldots + a_n\varepsilon^{n-k} + \eta_n\varepsilon^{n-k}$$

Then $\eta_k \simeq 0$. One has the following Goze decomposition

$$\binom{\varepsilon}{\eta_{k-1}} = \varepsilon\binom{1}{a_k} + \varepsilon\eta_k\binom{0}{1}$$

So $\binom{\varepsilon}{\eta_{k-1}} = T_k\binom{\varepsilon}{\eta_k}$, where T_k is the standard quadratic transformation $(X,Y) \to (X,(a_k+Y)X)$.

Secondly, suppose there are infinitesimals $\eta_0 = \eta, \eta_1, \ldots, \eta_n$ and standard numbers $a_1,\ldots,a_n$ such that

$$\binom{\varepsilon}{\eta_{k-1}} = \begin{pmatrix} 1 & 0 \\ a_k & \varepsilon \end{pmatrix}\binom{\varepsilon}{\eta_k} \qquad 1 \leq k \leq n$$

Then

$$\binom{\varepsilon}{\eta} = \begin{pmatrix} 1 & 0 \\ a_1 + a_2\varepsilon + \ldots + a_n\varepsilon^{n-1} & \varepsilon \end{pmatrix}\binom{\varepsilon}{\eta_n}$$

Hence $\eta = a_1\varepsilon + a_2\varepsilon^2 + \ldots + a_n\varepsilon^n + \eta_n\varepsilon^n$.

Finally, we present a possible way to turn every Goze-decomposition of an infinitesimal vector $\binom{\varepsilon}{\eta}$ into a shadow expansion of ε and of η. Assume one has a decomposition

$$\binom{\varepsilon}{\eta} = (U_1\ \alpha_1\ V_1)(U_2\ \alpha_2\ V_2)\dots(U_{n-1}\ \alpha_{n-2}\ V_{n-2})\binom{\alpha_{n-1}}{\alpha_n} \qquad \text{(st } n)$$

Where $U_{n-1}, U_{n-2}, V_1, \dots, V_{n-2}$ are standard vectors and $\alpha_1, \dots, \alpha_n \simeq 0$. Then $\varepsilon = P(\alpha_1, \dots, \alpha_n)$ and $\eta = Q(\alpha_1, \dots, \alpha_n)$, where P and Q are polynomials in $\alpha_1, \dots, \alpha_n$ with standard coefficients. By symmetry, it suffices to expand P. Let us write

$$\varepsilon = b_1\pi_1 + \dots + b_k\pi_k \qquad \text{(st } k)$$

where every π_j is a product of the α_i, and st b_j $(i \leq j \leq k)$. Now determine a Goze decomposition of the vector $(\pi_1, \dots, \pi_k)^T$. There are standard vectors $W_i = (w_{1i}, \dots, w_{ki})^T$ and infinitesimals $\delta_1, \dots, \delta_k$ such that

$$\begin{pmatrix} \pi_1 \\ \vdots \\ \pi_k \end{pmatrix} = \delta_1 W_1 + \dots + \delta_1 \cdot \dots \cdot \delta_k W_k$$

Put $c_i = b_1 w_{1i} + \dots + b_k w_{ki}$ $(1 \leq i \leq k)$. Then st c_i $(1 \leq i \leq k)$ and

$$\varepsilon = c_1\delta_1 + \dots + c_k\delta_1 \cdot \dots \cdot \delta_k$$

is a shadow expansion of ε (with respect to the scale $(\delta_1, \dots, \delta_1 \cdot \dots \cdot \delta_k)$).

4. Shadow expansions and approximations.

We now study the way numbers a having some shadow expansion $\sum_{n=0} c_n u_n$ are approximated by the partial sums of the series $\sum_{n=0} c_n u_n$.

Notation: Let $(u_n)_{n\in\mathbb{N}}$ be an order scale and $a \sim_n \sum_{n=0} c_n u_n$. Let $n \in \mathbb{N}$. The term $c_n u_n$ will be written T_n and the remainder $a \sim_n \sum_{k=0} T_k$ will be written R_n. We write $k+p(k)$ the smallest index larger than k such that $T_{k+p(k)} \neq 0$ (if such an index exists). See Definition 2.23.

The proof of the following proposition is elementary.

Proposition 6.17: Let $(u_n)_{n\in\mathbb{N}}$ be an order scale and $(c_n)_{n\in\mathbb{N}}$ be a standard sequence. Let $a \sim \sum_{n=0} c_n u_n$. Then for all st $n \in \mathbb{N}$.

(i) if $n+p(n)$ is not defined, then $R_n \in u\text{-}M$

(ii) if $n+p(n)$ is defined, then $R_n = (1+\emptyset)T_{n+p(n)}$ and $R_{n+p(n)}/R_n \simeq 0$.

Next theorem resumes some fundamental characteristics of shadow expansions with respect to approximation. It is a direct consequence of the previous proposition and the representation $a = S_p + \mu$ (see page 161).

Theorem 6.18: Let $(u_n)_{n\in\mathbb{N}}$ be an order scale and let $a \sim \sum_{n=0} c_n u_n$ be a shadow expansion. Suppose a new non-zero term is added to a partial sum of standard degree of the series $\sum_{n=0} c_n u_n$. Then the relative precision increases with an unlimited amount.

There are indices such that the precision belongs to the microhalo of the scale $(u_n)_{n\in\mathbb{N}}$.

Without further analysis, nothing more can be said about the approximation of a number a by the partial sums of its shadow expansion, and in particular about the index of best approximation. In some cases it is possible to get information from an analysis of the approximation factor $A_n = R_{n-1}/T_n$. It may also be possible to compare the number a with some other number, which has the same shadow expansion, and which is approximated by the partial sums in a known manner. In Chapter II and III we were able to follow this strategy.

In this chapter and the previous section we considered formal expansions with respect to some scale $(u_n)_{n\in\mathbb{N}}$ and deduced a universal characteristic of the rapidity of approximation: if $(R_n)_{n\in\mathbb{N}}$ is the sequence of the remainders associated to the expansion $a \sim \sum_{k=0} c_k u_k$, then "$R_n \in$ u-M for all unlimited n up to some unlimited index ω".

As far as approximations are considered, one might consider the reciprocal: one could take the rapidity of approximation as a starting point, and look for sequences satisfying this rate.

In this sense, the condition "$R_n \in$ u-M for all unlimited n up to some unlimited index ω" means a weakening of the condition "a possesses a shadow expansion with respect to the scale $(u_n)_{n\in\mathbb{N}}$".

This manner of stressing the rapidity of an approximation, rather than formal developments, might be a way to include the polynomials in 1/z associated to the Besselfunction $J_p(z)$ into our theory. Indeed, we have the divergent expansion (see Chapter I.3):

$$J_p(z) = \sum_{k=0} \sqrt{\frac{2}{\pi}}\,\frac{(\frac{1}{4}-p^2)(\frac{9}{4}-p^2).\ \dots\ .((k-\frac{1}{2})^2-p^2)}{k!2^k}.\cos(z-\frac{\pi}{4}-\frac{k}{2}\pi).\frac{1}{z^{k+1/2}}$$

Let z be positive and unlimited. Then the above expansion of $J_p(z)$ cannot be a shadow expansion in the case that z mod(2π) is nonstandard.

But it follows from the estimation of the remainders (Theorem 1.8(ii)) that

$$|R_{n-1}(z)| \leq \left| 2.\sqrt{\tfrac{2}{\pi}}\,\frac{(\frac{1}{4}-p^2)(\frac{9}{4}-p^2).\ \dots\ .((n-\frac{1}{2})^2-p^2)}{n!2^n}.\frac{1}{z^{n+1/2}}\right|$$

Hence the approximation of $J_p(z)$ by the partial sums of the above expansion satisfies the criterium on rapidity of approximation of shadow expansions with respect to the order scale $(1/z^{n+\frac{1}{2}})_{n\in\mathbb{N}}$, namely "$R_n \in 1/z\text{-}M$ for all unlimited indices n up to some unlimited index ω".

Let us briefly consider some other rates of approximation. The property "$R_n \in hal(0)$ for all unlimited n up to some unlimited index ω" is satisfied by approximations of standard numbers by terms of standard sequences converging to it. But also by the approximation of $e^{1+\varepsilon}$ by the partial sums $\sum_{k=0}^{n} \frac{1}{k!}$ or even $\sum_{k=0}^{n} \frac{1+\varepsilon k^k}{k!}$ $(\varepsilon \simeq 0)$. The property "$R_0 \in hal(0)$" is satisfied by the approximation of a limited number by its shadow.

5. "Summation to the smallest term".

Consider a convergent series whose terms tend monotonously to 0. Clearly the number which is best approximated by the sequence of partial sums of the series is its sum. Now consider a divergent series $\sum_{n=0} c_n u_n$ whose terms first decrease in absolute value, assume a minimal value for some index, say K, and after that start to increase -like the series $\sum_{n=0} (-1)^n n!\varepsilon^n$ studied in the first chapter of this book-. By analogy, one could consider the numbers which are most close to $\sum_{n=0}^{K} c_n u_n$ as best approximated by the sequence of partial sums of this series.

Let us state the problem in a different way. Given a number a such that $a \sim \sum_{n=0} c_n u_n$, is it true that a is best approximated by the partial sum $\sum_{n=0}^{K} c_n u_n$? This problem of "summation to the smallest term" had already been considered by Poincaré (for divergent asymptotic expansions). In many cases, the remainders are less than the first neglected term (see for instance [54] or [63]). Then the property of summation to the smallest term is true for the majorations of the remainders.

In Chapter III we considered cases -divergent asymptotic expansions of a certain class of integral transformations- where summation to the smallest term indeed provided nearly-optimal approximation results. We did so by studying the approximation factor $A_n(\varepsilon)$, i.e. the quotient between the existing remainder $R_{n-1}(\varepsilon)$ and the first neglected term $T_n(\varepsilon)$.

Inspection of the argument shows that for the class of analytic functions under consideration, the approximation factor satisfies a particular property, i.e.

whenever $\varepsilon \simeq 0$, then $A_n(\varepsilon) \simeq A_{n+1}(\varepsilon)$ for all n. Below we show that this property alone suffices to establish the property of summation to the smallest term.
Of course the property "R_K is nearly minimal" is not always verified. Even the much weaker property $\sum_{n=0}^{K} c_n u_n \sim \sum_{n=0} c_n u_n$ is not always true; in other words, when the shadow expansion of a number is truncated at the index of the smallest term, there is no guarantee that the resulting error belongs to the u-microhalo. This and related problems are studied in the second half of this section.

Notation: As in Definition 3.1, the lowest index of the minimal terms of an expansion is written K.

Definition 6.19: (i) A sequence $(a_n)_{n\in\mathbb{N}}$ of real numbers is called a small-step sequence if $a_{n+1} \simeq a_n$ for all $n \in \mathbb{N}$ (see also [3]).
(ii) A sequence $(a_n)_{n\in\mathbb{N}}$ is called S-increasing if $n < m$ and $a_n \not\simeq a_m$ implies $a_n < a_m$.

Note that the sequences of approximation factors $A_n(\varepsilon)$ associated to the divergent expansions of the Borel and Laplace transforms of Chapter III were S-decreasing small-step sequences (for $\varepsilon > 0$). Approximations of the Borel and Laplace transforms based on these expansions satisfy the property of summation to the (nearly) smallest term; this was proved by direct verification. In some sense, next theorem generalizes this result:

Theorem 6.20: Let $a \sim \sum_{n=0} c_n \varepsilon^n$ where $\varepsilon \simeq 0$, $\varepsilon > 0$, $(c_n)_{n\in\mathbb{N}}$ is a standard sequence of nonzero real numbers and $\sum_{n=0} c_n \varepsilon^n$ is divergent. Let T_n, R_n, A_n and K be as above. Suppose $(|T_{n+1}/T_n|)_{n\in\mathbb{N}}$ is S-increasing and A_n is a small step sequence. Then $|R_{K-1}|$ is nearly minimal as far as indices n such that T_{n+1}/T_n is limited are concerned.

The theorem will be a consequence of the following lemma.

Lemma: (i) For all $n \in \mathbb{N}$ one has (1) $A_n = \dfrac{1+\emptyset . T_{n+1}/T_n}{1-T_{n+1}/T_n}$

(ii) As long as T_{n+1}/T_n is appreciable, the terms T_n are alternating.

(iii) As long as T_{n+1}/T_n is limited, the following formulas hold.

(2) $A_n = \dfrac{1+\emptyset}{1-T_{n+1}/T_n}$ (3) $\dfrac{T_{n+1}}{T_n} = \dfrac{A_n - 1}{A_n + \emptyset}$ (4) $\dfrac{R_n}{R_{n-1}} = (1+\emptyset)\dfrac{T_{n+1}}{T_n}$

Proof: (i) The equality $R_{n-1} = R_n + T_n$ implies $A_n T_n = A_{n+1} T_{n+1} + T_n$. Hence

$$A_n = \frac{1+\emptyset . T_{n+1}/T_n}{1-T_{n+1}/T_n}$$

(ii) For all st n one has $A_n \simeq 1$. Now $|T_{n+1}/T_n| \to \infty$, so by formula (1) there exists n such that $A_n \not\simeq 1$. Suppose $A_n \gtrapprox 1$. Because A_n is a small step sequence, there exists $\bar{n}$ such that $A_{\bar{n}} \gtrapprox 1$ is limited. Then $1 \gtrapprox T_{\bar{n}+1}/T_{\bar{n}} \gtrapprox 0$. Again by formula (1) for $n > \bar{n}$ one must have $1 > T_{n+1}/T_n \gtrsim T_{\bar{n}+1}/T_{\bar{n}}$, for $(|T_{n+1}/T_n|)_{n\in\mathbb{N}}$ is S-monotonous, and A_n is a small step sequence. But this contradicts the fact that $|T_{n+1}/T_n| \to \infty$. So if A_n is no longer equivalent to 1, one must have $A_n \lessapprox 1$, which implies that $T_{n+1}/T_n \lessapprox 0$. Because A_n is a small step sequence the property $A_n \lessapprox 1$ must hold up to some $\tilde{n}$ such that $A_{\tilde{n}} \simeq 0$. This implies that $T_{n+1}/T_n \lessapprox 0$ at least as long as T_{n+1}/T_n is appreciable.

(iii) The formulas result from easy calculations.

<u>Proof of the theorem</u>: By lemma (iii), formula (4), the remainder can only be minimal for one of the indices n such that $|T_{n+1}/T_n| \simeq 1$. By lemma (ii) this implies $T_{n+1}/T_n \simeq -1$. Then $A_n \simeq 1/2$ by formula (2) and $R_{n-1} = (1+\emptyset)T_n/2$ by the definition of A_n. So for all n such that $T_{n+1}/T_n \simeq -1$ one has

$$\left|\frac{R_{K-1}}{R_{n-1}}\right| = (1+\emptyset)\left|\frac{T_K}{T_n}\right| \lessapprox 1$$

Hence R_{K-1} is nearly minimal as long as T_{n+1}/T_n is limited.

Suppose that the index n is so large that T_{n+1}/T_n fails to be limited. Then $A_n \simeq 0$, so R_{n-1} and T_n are no longer "linked". This implies that R_{n-1} might be very small "by accident". This situation is excluded if we demand that $A_{n+1} = (1+\emptyset)A_n$ for all $n \in \mathbb{N}$ (then A_{n+1}/A_n is a small step sequence). With this condition added to the conditions of Theorem 6.20 it is easily verified that R_{K-1} is nearly minimal with respect to all remainders. Note that the property $A_{n+1} = (1+\emptyset)A_n$ is verified in the case of the Borel and Laplace transforms treated in Chapter III.

We now discuss the question whether always $\sum_{n=0}^{K} c_n u_n \sim \sum_{n=0} c_n u_n$. The propositions and examples below suggest that in practice the answer is positive, and that exceptions are only to be found between very slowly divergent series. Proposition 6.24 states such an exception for expansions in ε-shadow.

<u>Notation</u>: We maintain the notation "n+p(n)" for indices as defined in the preceding section. Let $(u_n)_{n\in\mathbb{N}}$ be an order scale and $(c_n)_{n\in\mathbb{N}}$ be a standard sequence. As usual let ρ be a universal index associated to this scale and u-M be its microhalo

We write M the first index of a relative minimum in the sequence $|c_n u_n|$, i.e.

$$|c_n u_n| > |c_M u_M| \quad \text{for all } n < M \text{ such that } c_n \neq 0$$
$$\exists k \text{ such that } |c_M u_M| = |c_{M+p(M)} u_{M+p(M)}| = \ldots = |c_k u_k| < |c_{k+p(k)} u_{k+p(k)}|$$

For $\lambda > 1$ we define M_λ to be the index such that

$$|T_{n+p(n)}/T_n| < 1-1/\lambda \quad \text{for all } n \text{ such that } n+p(n) \leq M_\lambda$$
$$|T_{M_\lambda+p(M_\lambda)}/T_{M_\lambda}| \geq 1-1/\lambda$$

<u>Proposition 6.21</u>: (summation nearly to the minimal term). Let $(u_n)_{n\in\mathbb{N}}$ be an order scale and $(c_n)_{n\in\mathbb{N}}$ be a standard sequence. Let $\lambda \gnsim 1$ be an asymptotically limited number with respect to the scale $(u_n)_{n\in\mathbb{N}}$. Then $\sum_{n=0}^{M_\lambda} c_n u_n \sim \sum_{n=0} c_n u_n$.

<u>Proof</u>: Assume $M_\lambda > \rho$, otherwise nothing needs to be proved. We will show that $\sum_{n=\rho+1}^{M_\lambda} \in M$. Indeed

$$\left|\sum_{n=\rho+1}^{M_\lambda} c_n u_n\right| \leq |c_{\rho+p(\rho)} u_{\rho+p(\rho)}| \sum_{n=\rho+1}^{M_\lambda} (1-1/\lambda)^{n-\rho+1}$$

$$\leq |c_{\rho+p(\rho)} u_{\rho+p(\rho)}| \sum_{m=0}^{\infty} (1-1/\lambda)^m$$

$$= |c_{\rho+p(\rho)} u_{\rho+p(\rho)}| . \lambda \in u\text{-}M.$$

Hence $\sum_{n=0}^{M_\lambda} c_n u_n \sim \sum_{n=0} c_n u_n$.

<u>Proposition 6.22</u>: Let $(u_n)_{n\in\mathbb{N}}$ be an order scale and $(c_n)_{n\in\mathbb{N}}$ be a standard alternating sequence. Then $\sum_{n=0}^{M} c_n u_n \sim \sum_{n=0} c_n u_n$.

<u>Proof</u>: Assume $M > \rho$, otherwise nothing needs to be proved. Let us write

$$\sum_{n=0}^{M} c_n u_n = \sum_{n=0}^{\rho} c_n u_n + \sum_{n=\rho+1}^{M} c_n u_n$$

Because $(c_n u_n)_{n\in\mathbb{N}}$ is alternating and decreasing up to M, one has

$$\left|\sum_{n=\rho+1}^{M} c_n u_n\right| \leq |c_{\rho+p(\rho)} u_{\rho+p(\rho)}| \in u\text{-}M$$

Hence $\sum_{n=0}^{M} c_n u_n \sim \sum_{n=0} c_n u_n$.

<u>Proposition 6.23</u>: Let $(u_n)_{n\in\mathbb{N}}$ be an order scale and $(c_n)_{n\in\mathbb{N}}$ be a standard sequence. Suppose M is asymptotically limited with respect to the scale $(u_n)_{n\in\mathbb{N}}$. Then $\sum_{n=0}^{M} c_n u_n \sim \sum_{n=0} c_n u_n$.

Proof: Write

$$\sum_{n=0}^{M} c_n u_n = \sum_{n=0}^{\rho} c_n u_n + \sum_{n=\rho+1}^{M} c_n u_n$$

Now $\left|\sum_{n=\rho+1}^{M} c_n u_n\right| \leqq Mc_{\rho+p(\rho)} u_{\rho+p(\rho)} \in u\text{-}M$. Hence $\sum_{n=0}^{M} c_n u_n \sim \sum_{n=0} c_n u_n$.

Examples: We consider expansions in $1/\omega$-shadow with ω positive unlimited.

1) Let st $k > 0$ and put $a_n = (n!)^{1/k}$. Then the (first) index of the smallest term of the series $\sum_{n=0} a_n/\omega^n$ equals $K = [\omega^k]-1$. So K is asymptotically limited with respect to the scale $(1/\omega^n)_{n\in\mathbb{N}}$. Hence $\sum_{n=0}^{K} a_n/\omega^n \sim \sum_{n=0} a_n/\omega^n$.

Now put $b_n = n^{n/k}$. Here $K = (1+\emptyset)(\omega/e)^k$, so again K is asymptotically limited. Hence $\sum_{n=0}^{K} b_n/\omega^n \sim \sum_{n=0} b_n/\omega^n$.

2) The estimations figuring in the proofs of propositions 6.21, 6.22 and 6.23 were very rough. Here is a series $\sum_{n=0} c_n/\omega^n$ which does not satisfy the conditions of any of these propositions, but for which the property $\sum_{n=0}^{K} c_n/\omega^n \sim \sum_{n=0} c_n/\omega^n$ still holds. We write $\log_p(x)$ for the pth iterative of the logarithmical function and $e_p(x)$ for the pth iterative of the exponential function. Put

$$c_n = \log_p(e_p(1)+1).\log_p(e_p(1)+2).\ \ldots\ .\log_p(n).$$

The index of the (first) smallest term of this series equals $K = [e_p(\omega)](-1)$. Now K is neither asymptotically limited with respect to the scale $(1/\omega^n)_{n\in\mathbb{N}}$, nor $1-T_K/T_{K-1} \in 1/\omega\text{-}M$, for

$$T_K/T_{K-1} \geqq \frac{\log_p(e_p(\omega)-1)}{\omega} = 1-(1+\emptyset)\frac{e^{-e_{p-1}(\omega)-e_{p-2}(\omega)-\ \ldots\ -\omega}}{\omega}$$

However,

$$\frac{T_{[e_{p-1}(\omega)]+1}}{T_{[e_{p-1}(\omega)]}} \leqq \frac{\log_p(e_{p-1}(\omega))}{\omega} = \frac{\log\omega}{\omega} \simeq 0$$

Hence, by proposition 6.21,

$$\sum_{n=e_p(1)}^{[e_{p-1}(\omega)]} c_n/\omega^n \sim \sum_{n=e_p(1)} c_n/\omega^n$$

Furthermore,

$$\sum_{n=[e_{p-1}(\omega)]+1}^{[e_p(\omega)]} c_n/\omega^n \leq e_p(\omega).T_{[e_{p-1}(\omega)]}$$

$$\leq e^{e_{p-1}(\omega)} \left(\frac{\log(\omega)}{\omega}\right)^{e_{p-1}(\omega)}$$

$$\leq \left(\frac{1}{\sqrt{\omega}}\right)^{e_{p-1}(\omega)} \in 1/\omega\text{-M}$$

Hence $\sum_{n=e_p(1)}^{K} c_n/\omega^n \sim \sum_{n=e_p(1)} c_n/\omega^n$.

The last proposition shows that not always $\sum_{n=0}^{K} a_n/\omega^n \sim \sum_{n=0} a_n/\omega^n$.

Proposition 6.24: Let $\omega \in \mathbb{N}$ be unlimited. There exists a standard sequence $(a_n)_{n\in\mathbb{N}}$ such that $\sum_{n=0}^{K} a_n/\omega^n$ does not have the expansion in $1/\omega$-shadow $\sum_{n=0}^{n} a_n/\omega^n$.

Proof: We recall the definition of $n \uparrow p$.

$$0 \uparrow p = 1$$

$$n \uparrow p = n^{n^{\cdot^{\cdot^{n}}}} \ \text{p times} \qquad n > 0$$

Define a sequence $(q_k)_{k\in\mathbb{N}}$ by

$$q_0, q_1, q_2, q_3 = 1$$

$$q_k = m - \frac{1}{m \uparrow m} \qquad \text{if} \quad m \uparrow m \leq k < (m+1) \uparrow (m+1) \qquad (k \geq 4)$$

Finally, put $a_n = \prod_{k\leq n} q_k$. The series $\sum_{n=0} a_n/\omega^n$ is divergent, and the index of the smallest term equals $K = (\omega+1) \uparrow (\omega+1)$. Let us write the following identity:

$$\sum_{n=0}^{(\omega+1)\uparrow(\omega+1)} a_n/\omega^n = \sum_{n=0}^{\omega\uparrow\omega-1} a_n/\omega^n + \sum_{n=\omega\uparrow\omega}^{(\omega+1)\uparrow(\omega+1)} a_n/\omega^n + a_{(\omega+1)\uparrow(\omega+1)}/\omega^{(\omega+1)\uparrow(\omega+1)}$$

Because the terms a_n/ω^n are decreasing, the first sum on the right side of the equality is minorized by $\omega\uparrow\omega . a_{\omega\uparrow\omega}/\omega^{\omega\uparrow\omega}$. Concerning the second sum, let us write

$$\sum_{n=\omega\uparrow\omega}^{(\omega+1)\uparrow(\omega+1)_{-1}} a_n/\omega^n = (a_{\omega\uparrow\omega}/\omega^{\omega\uparrow\omega}) \sum_{n=\omega\uparrow\omega}^{(\omega+1)\uparrow(\omega+1)_{-1}} \frac{a_n}{a_{\omega\uparrow\omega}} . \frac{1}{\omega^{n-\omega\uparrow\omega}}$$

For $\omega\uparrow\omega \leq n < (\omega+1) \uparrow (\omega+1)$ one has $\frac{a_n}{a_{\omega\uparrow\omega}} = \left(\omega - \frac{1}{(\omega+1)\uparrow(\omega+1)}\right)^{n-\omega\uparrow\omega}$. Hence

$$\sum_{n=\omega\uparrow\omega}^{(\omega+1)\uparrow(\omega+1)-1} \frac{a_n}{a_{\omega\uparrow\omega}}\cdot\frac{1}{\omega^{n-\omega\uparrow\omega}} \geqq ((\omega+1)\uparrow(\omega+1)-\omega\uparrow\omega)\left(1-\frac{1}{\omega(\omega+1)\uparrow(\omega+1)}\right)^{(\omega+1)\uparrow(\omega+1)-\omega\uparrow\omega}$$

$$\geqq \frac{(\omega+1)\uparrow(\omega+1)-\omega\uparrow\omega}{2}$$

Combining,

$$\sum_{n=0}^{(\omega+1)\uparrow(\omega+1)} a_n/\omega^n \geqq \frac{1}{2}\left(\frac{(\omega+1)\uparrow(\omega+1)\cdot a_{\omega\uparrow\omega}}{\omega^{\omega\uparrow\omega}}\right)$$

$$\geqq \frac{a_{\omega\uparrow\omega}}{2}$$

Hence $\sum_{n=0}^{K} a_n/\omega^n$ is unlimited, and cannot be expanded in $1/\omega$-shadow.

References.

I Introductions to Internal Set Theory

1. F. Diener, Cours d'analyse nonstandard, Office des Publ. Univ., Alger, 1983.
2. F. Diener, G. Reeb, Analyse Nonstandard, Hermann, Paris (to appear).
3. M. Diener, Une initiation aux outils fondamentaux de l'analyse nonstandard, in: Ecole d'été en analyse nonstandard et représentation du réel, Oran 1984, OPU, Alger, éd. CNRS, Paris, 1985.
4. R. Lutz, M. Goze, Nonstandard analysis. A practical guide with applications, Springer Lecture Notes in Mathematics 881, 1981.
5. E. Nelson, Internal Set Theory, Bull. Amer. Math. Soc. 83, p. 1165-1198, 1977.
6. E. Nelson, The syntax of nonstandard analysis, Ann. of Pure and App. Logic. (to appear).
7. M.K. Richter, Ideale Punkte, Monaden und Nichtstandard-Methoden, Vieweg, 1982.
8. A. Robert, Analyse nonstandard, Presses Polytechniques romandes, Lausanne, 1985.

II Other references, nonstandard

9. E. Benoit, Systèmes lent-rapides dans $\mathbb{R}^3$ etleurs canards, Astérisque 110, P 159-191, 1983.
10. E. Benoit, Canards de $\mathbb{R}^3$, Thèse, Univ. de Nice, 1984.
11. E. Benoit, J.-L. Callot, F. and M. Diener, Chasse au canard, Collectanea Mathematica, 31-3, 1980.
12. I.P. van den Berg, Approximations asymptotiques et ensembles externes, Thèse, Publ. IRMA, Strasbourg, 1984.
13. I.P. van den Berg, Sur la sommation au plus petit terme des séries divergentes in: Mathématiques finitaires et analyse nonstandard, Luminy 1985 (to appear).
14. I.P. van den Berg, Un point de vue nonstandard sur les développements en série de Taylor, Astérisque 110, p. 209-223, 1983.
15. I.P. van den Berg, Un principe de permanence général, Asterisque 110, p. 193-208, 1983.
16. I.P. van den Berg, M. Diener, Diverses applications du lemme de Robinson en analyse nonstandard, C.R. Ac. Sci. Paris, sér. I, 293, p.501-504, 1981.
17. M. Canalis Durand, F. Diener, M. Gaetano, Calcul des valeurs à canard à l'aide de MACSYMA, in: Mathématiques finitaires et analyse nonstandard, Luminy 1985 (to appear).
18. F. Diener, Développements en ε-ombres, Outils et modèles mathématiques pour l'automatique, l'analyse des systèmes et le traitement du signal, T. 3, éd. CNRS p. 315-328, 1983.

19. F. Diener, Fleuves et variétés centrales, in: Actes Journées S.M.F. "Singularités d'Equations Différentielles", Astérisque (to appear).

20. F. Diener, Propriétés asymptotiques des fleuves, C.R. Ac. Sci. Paris, sér. I, 302, p. 55-58, 1986.

21. F. Diener, Sauts des équations $\varepsilon\ddot{x} = f(t,x,\dot{x})$, SIAM J. Math. Anal., Vol. 17, no 3, p. 533-559, 1986.

22. M. Diener, Détermination et existence des fleuves en dimension 2, C.R. Ac. Sci. Paris, sér. I, 301, p. 899-902, 1985.

23. M. Diener, Etude générique des canards, Thèse, Publ. IRMA, Strasbourg, 1981.

24. M. Diener, The canard unchained, or how fast/slow dynamical systems bifurcate, Math. Intell. 6.3, p. 38-49, 1984.

25. M. Diener, I.P. van den Berg, Halos et Galaxies, C.R. Ac. Sci. Paris, Sér. I, 293, p. 385-388, 1981.

26. M. Diener, G. Reeb, Champs polynomiaux: nouvelles trajectoires remarquables, Bulletin de l'Acad. Royale de Belgique (to appear).

27. M. Goze, Etude locale des courbes algébriques planes, Astérisque 110, p. 245-259, 1983.

28. M. Goze, J.M. Ancochea Bermudez, Algèbres de Lie rigides, Indag. Math. Vol. 47, Fasc 4, p. 397-415, 1985.

29. K. Hrbacek, Nonstandard Set Theory, Am. Math. Monthly 85, no. 8, p. 659-677, 1979.

30. T. Ismael, Etude macrocoscopique de certaines courbes algébriques, Thèse 3e cycle, Publ. IRMA, Strasbourg, 1984.

31. I. Lakatos, Cauchy and the continuum, in: Mathematics, science and epistemology, Vol 2, p. 43-60, Cambridge Univ. Press, 1978.

32. D. Laugwitz, Infinitesimalkalkül, Bibliographisches Institut, Mannheim, 1978.

33. D. Laugwitz, An Abraham Robinson's sequential lemma, preprint, 1974.

34. A.H. Lightstone, A. Robinson, Non-Archimedian fields and asymptotic expansions, North-Holland, Amsterdam, 1975.

35. A. Robinson, Nonstandard analysis, North-Holland, Amsterdam, 1966, 2nd ed. 1973.

36. K. Stroyan, W.A.J. Luxemburg, Introduction to the theory of infinitesimals, Acad. Press, New York, 1976.

37. A.K. Zvonkin, M.A. Shubin, Nonstandard analysis and singular perturbations of ordinary differential equations, Russian Math. Surveys 39.2, p. 69-131, 1984.

III Other references, classical

38. J.R. Airey, The "converging factor" in asymptotic series and the calculation of Bessel, Laguerre and other functions, Philos. Mag. 24, p. 521-552, 1937.

39. L. Berg, Zur Abschätzung des Restgliedes in der asymptotischen Entwicklung des Exponentialintegrals, Computing 18, p. 361-363, 1977.

40. J.G. van der Corput, Asymptotic developments I, Fundamental theorems of asymptotics, J. d'Analyse Mathématique 4, p. 341-418, 1956.

41. J.G. van der Corput, Introduction to the neutrix calculus, MRC Technical Summer report, 128-130, Madison,Wisconsin, 1960.

42. P. Dienes, The Taylor series, Clarendon Press, Oxford, 1931.

43. R.B. Dingle, Asymptotic expansions; their derivation and interpretation, Acad. Press, London, 1973.

44. W. Eckhaus, Asymptotic analysis of singular perturbations, North-Holland, Amsterdam, 1979.

45. L. Gillman, M. Jerison, Rings of continuous functions, Springer Verlag, 1960, 2nd ed. 1980.

46. G.H. Hardy, Orders of infinity, the "Infinitärcalcul" of Paul du Bois-Reymond, Cambridge Univ. Press, 1910.

47. P. Henrici, Applied and computational complex analysis, vol. 2, Wiley & Sons, New York, 1977.

48. M. Holtz, Über das Restglied asymptotischer Entwicklungen von Laplace integralen für $s \to \infty$, Z. für Ang. Math. und Mech. 48, p. 131-134, 1968.

49. H. Jeffreys, The remainder in Watson's lemma, Proc. Roy. Soc. London, Ser. A, 248, p. 88-92, 1958.

50. H.M. Kammerer, Sine and Cosine approximation curves, Amer. Math. Monthly 43, p. 293-294, 1936.

51. H.A. Lauwerier, Asymptotic analysis, Math. Centre Tracts 54, Amsterdam, 1977.

52. J. Martinet, J.-P. Ramis, "Problèmes de modules pour des équations différentielles non linéares du premier ordre, Publ. Math. I.H.E.S. 55, p. 63-164, 1982.

53. F. Nevanlinna, Zur Theorie der Asymptotischen Potenzreihen, Ann. Ac.Sc. Fennicae, ser. A, Fom XII, Helsinki, 1919.

54. F.W.J. Olver, Asymptotics and special functions, Acad. Press, New York, 1974.

55. H. Poincaré, Les méthodes nouvelles de la mécanique céleste, T. II, 1893, (also: Dover).

56. H. Poincaré, Sur les integrales irrégulières des équations linéares, Acta Math. 8, p. 295-344, 1886.

57. S. Priess-Crampe, Angeordnete Strukturen, Gruppen, Körper, Projektive Ebenen, Springer Verlag, 1983.

58. J.-P. Ramis, Les séries k-sommables et leurs applications, Springer Lecture Notes in Physics, 126, p. 178-199, 1980.

59. J.-P. Ramis, Théorèmes d'indices Gevrey pour les équations differentielles ordinaires, Mem. of the AMS no 296, 1984.

60. H. Schell, Eine asymptotische Aussage über das absolut kleinste Restgebied von alternierenden asymptotischen Entwicklungen, Math. Nachr. 44, p. 157-164, 1970.

61. Th.J. Stieltjes, Recherches sur quelques séries sémi-convergentes, Ann. Sc. Ecol. Norm. Sup. 3, p. 201-258, 1886.

62. G.N. Watson, A treatise on the theory of Bessel functions, Cambridge Univ. Press, 2nd ed., London, 1944.

63. R. Wong, Error bounds for asymptotic expansions of integrals, SIAM Review, vol. 22, no 4, 1984.

Index/Lexicon

Let X be an external set, containing only standard elements. Let f be an external function defined on X, with internal values. The extension principle says that f can be extended to an internal function defined on some internal set $I \supsetneq X$. It is a consequence of the so-called Saturation principle, proved in [6].

Let E be a formula of IST such that E(0) and $(\forall^{st} n \in \mathbb{N})(E(n) \rightarrow E(n+1))$ hold. The external induction principle says that then E(n) for all st $n \in \mathbb{N}$.

Idealization axiom

This is one of the axioms of Internal Set Theory. It runs as follows. Let B(x,y) be an internal formula. Let $\forall^{st\ fin}z$ mean $(\forall z)(\mathrm{st}\ z \wedge z \text{ finite} \rightarrow \ldots)$ Then

$$\forall^{st\ fin} z\ \exists y\ \forall x \in z\ B(x,y) \Leftrightarrow \exists y (\forall^{st} x) B(x,y)$$

Let us restate the nontrivial part of this equivalence in words: if for all standard finite sample $x_0,\ldots,x_k$ there exists y such that the formula's $B(x_0,y),\ldots,B(x_k,y)$ all hold, then there exists $\overline{y}$ such that the formula's $B(x,\overline{y})$ hold for all standard x. We made full use of this axiom to prove the general permanence principle. One of the main consequences of this axiom is that every infinite standard set, in particular $\mathbb{N}$ and $\mathbb{R}$, contains nonstandard elements.

Infinitesimal

Infinitesimal numbers, or infinitesimals, are real numbers which in absolute value are smaller than all standard strictly positive real numbers. The existence of infinitesimals other than 0 is a consequence of the Idealization axiom. If x-y is infinitesimal we write $x \simeq y$.

Internal: see Internal Set Theory

Internal Set Theory

Internal Set Theory (IST) is an extension of classical axiomatic set theory (Zermelo-Frankel set theory with axiom of choice ZFC). The only nonlogical symbol of ZFC is "$\in$". Here a new nonlogical symbol is added, the predicate "st" (standard), and three axioms to operate on it, Idealization, Standardization and Transfer. Formula's containing one or more of the symbols "ε" or "st" are called "formula's of IST". Formula's of IST which do not contain the symbol "st" are called internal, formula's of IST which do contain the symbol "st" are called external. We adopted the convention to call sets defined by an internal formula "internal" and sets defined by external formula's "external", only if they are not internal. IST is a conservative and relatively consistent extension of ZFC.

A limited real number is a number which in absolute value is smaller than some standard real number. In particular the limited and standard integers coincide.

The reduction algorithm is a mechanical way to replace a nonstandard theorem of IST by a classical theorem, which, within IST, is equivalent with it.

Limited real numbers x possess a shadow. It is the standard real number s such that $s \simeq x$. Its existence and uniqueness are a consequence of the Standardization and Transfer axioms. We write $s = {}^{o}x$. Likewise, vectors in $\mathbb{R}^n$, st n, possess a shadow: the standard vector whose components equal the shadows of the components of the original vector. The shadow ${}^{o}f$ of a real function $f: E \subset \mathbb{R} \to \mathbb{R}$ is only defined if f is of class S_0. It then equals

$${}^{o}f = {}^{S}(\mathrm{hal}\{(x,f(x)) \mid x \in E,\ x \text{ limited}\}$$

(it appears that if f is not of class S_0 the right-member of the above equality is not the graph of a function). The function ${}^{o}f$ may also be defined in two steps. First, define for all st $x \in E$

$$({}^{o}f)(x) = {}^{o}(f(x))$$

Secondly, extend f to ${}^{S}E$ by Standardization.

This is one of the axioms of Internal Set Theory. It runs as follows. Let C be any formula of IST. Then.

$$\forall^{st} x\ \exists^{st} y\ \forall^{st} z (z \in y \Leftrightarrow z \in x \wedge st\ z \wedge C(z))$$

In words: within a standard reference set x, the standard elements of x satisfying an arbitrary property C define a standard subset y of x. As a consequence of the Transfer axiom, this set y is unique. It is called the standardized of {st z ∈ x | C(z)}. We write

$$y = {}^{S}\{st\ z \in x \mid C(z)\}$$

The transfer principle is one of the three axioms of Internal Set Theory. Let A(x,t) be an internal formula, and let furthermore the parameter t be standard. Then

$$(\forall^{st} x)A(x,t) \Leftrightarrow (\forall x)A(x,t)$$

It is allowed that more parameters occur in A, but it must carefully be checked that they all have standard values (otherwise the application of the above rule may lead to serious mistakes). By its contraposition, the axiom implies that all explicitly defined sets of classical mathematics ($\emptyset$, 1, 2, 3, π, 10, 10^{10}, $\mathbb{N}$, $\mathbb{R}$,...) are standard. Indeed,

$$\exists! x\ A(x) \Rightarrow \exists^{st} x\ A(x)$$

Unlimited

Real numbers larger than any standard real number are called <u>positive unlimited</u>. Real numbers smaller than any standard real number are called <u>negative unlimited</u>. Their existence is a consequence of the Idealization Axiom.

Vol. 1090: Differential Geometry of Submanifolds. Proceedings, 1984. Edited by K. Kenmotsu. VI, 132 pages. 1984.

Vol. 1091: Multifunctions and Integrands. Proceedings, 1983. Edited by G. Salinetti. V, 234 pages. 1984.

Vol. 1092: Complete Intersections. Seminar, 1983. Edited by S. Greco and R. Strano. VII, 299 pages. 1984.

Vol. 1093: A. Prestel, Lectures on Formally Real Fields. XI, 125 pages. 1984.

Vol. 1094: Analyse Complexe. Proceedings, 1983. Edité par E. Amar, R. Gay et Nguyen Thanh Van. IX, 184 pages. 1984.

Vol. 1095: Stochastic Analysis and Applications. Proceedings, 1983. Edited by A. Truman and D. Williams. V, 199 pages. 1984.

Vol. 1096: Théorie du Potentiel. Proceedings, 1983. Edité par G. Mokobodzki et D. Pinchon. IX, 601 pages. 1984.

Vol. 1097: R. M. Dudley, H. Kunita, F. Ledrappier, École d'Éte de Probabilités de Saint-Flour XII – 1982. Edité par P.L. Hennequin. X, 396 pages. 1984.

Vol. 1098: Groups – Korea 1983. Proceedings. Edited by A.C. Kim and B.H. Neumann. VII, 183 pages. 1984.

Vol. 1099: C. M. Ringel, Tame Algebras and Integral Quadratic Forms. XIII, 376 pages. 1984.

Vol. 1100: V. Ivrii, Precise Spectral Asymptotics for Elliptic Operators Acting in Fiberings over Manifolds with Boundary. V, 237 pages. 1984.

Vol. 1101: V. Cossart, J. Giraud, U. Orbanz, Resolution of Surface Singularities. Seminar. VII, 132 pages. 1984.

Vol. 1102: A. Verona, Stratified Mappings – Structure and Triangulability. IX, 160 pages. 1984.

Vol. 1103: Models and Sets. Proceedings, Logic Colloquium, 1983, Part I. Edited by G. H. Müller and M. M. Richter. VIII, 484 pages. 1984.

Vol. 1104: Computation and Proof Theory. Proceedings, Logic Colloquium, 1983, Part II. Edited by M. M. Richter, E. Börger, W. Oberschelp, B. Schinzel and W. Thomas. VIII, 475 pages. 1984.

Vol. 1105: Rational Approximation and Interpolation. Proceedings, 1983. Edited by P.R. Graves-Morris, E.B. Saff and R.S. Varga. XII, 528 pages. 1984.

Vol. 1106: C.T. Chong, Techniques of Admissible Recursion Theory. IX, 214 pages. 1984.

Vol. 1107: Nonlinear Analysis and Optimization. Proceedings, 1982. Edited by C. Vinti. V, 224 pages. 1984.

Vol. 1108: Global Analysis – Studies and Applications I. Edited by Yu. G. Borisovich and Yu. E. Gliklikh. V, 301 pages. 1984.

Vol. 1109: Stochastic Aspects of Classical and Quantum Systems. Proceedings, 1983. Edited by S. Albeverio, P. Combe and M. Sirugue-Collin. IX, 227 pages. 1985.

Vol. 1110: R. Jajte, Strong Limit Theorems in Non-Commutative Probability. VI, 152 pages. 1985.

Vol. 1111: Arbeitstagung Bonn 1984. Proceedings. Edited by F. Hirzebruch, J. Schwermer and S. Suter. V, 481 pages. 1985.

Vol. 1112: Products of Conjugacy Classes in Groups. Edited by Z. Arad and M. Herzog. V, 244 pages. 1985.

Vol. 1113: P. Antosik, C. Swartz, Matrix Methods in Analysis. IV, 114 pages. 1985.

Vol. 1114: Zahlentheoretische Analysis. Seminar. Herausgegeben von E. Hlawka. V, 157 Seiten. 1985.

Vol. 1115: J. Moulin Ollagnier, Ergodic Theory and Statistical Mechanics. VI, 147 pages. 1985.

Vol. 1116: S. Stolz, Hochzusammenhängende Mannigfaltigkeiten und ihre Ränder. XXIII, 134 Seiten. 1985.

Vol. 1117: D.J. Aldous, J.A. Ibragimov, J. Jacod, Ecole d'Été de Probabilités de Saint-Flour XIII – 1983. Édité par P.L. Hennequin. IX, 409 pages. 1985.

Vol. 1118: Grossissements de filtrations: exemples et applications. Seminaire, 1982/83. Edité par Th. Jeulin et M. Yor. V, 315 pages. 1985.

Vol. 1119: Recent Mathematical Methods in Dynamic Programming. Proceedings, 1984. Edited by I. Capuzzo Dolcetta, W.H. Fleming and T. Zolezzi. VI, 202 pages. 1985.

Vol. 1120: K. Jarosz, Perturbations of Banach Algebras. V, 118 pages. 1985.

Vol. 1121: Singularities and Constructive Methods for Their Treatment. Proceedings, 1983. Edited by P. Grisvard, W. Wendland and J.R. Whiteman. IX, 346 pages. 1985.

Vol. 1122: Number Theory. Proceedings, 1984. Edited by K. Alladi. VII, 217 pages. 1985.

Vol. 1123: Séminaire de Probabilités XIX 1983/84. Proceedings. Edité par J. Azéma et M. Yor. IV, 504 pages. 1985.

Vol. 1124: Algebraic Geometry, Sitges (Barcelona) 1983. Proceedings. Edited by E. Casas-Alvero, G.E. Welters and S. Xambó-Descamps. XI, 416 pages. 1985.

Vol. 1125: Dynamical Systems and Bifurcations. Proceedings, 1984. Edited by B.L.J. Braaksma, H.W. Broer and F. Takens. V, 129 pages. 1985.

Vol. 1126: Algebraic and Geometric Topology. Proceedings, 1983. Edited by A. Ranicki, N. Levitt and F. Quinn. V, 423 pages. 1985.

Vol. 1127: Numerical Methods in Fluid Dynamics. Seminar. Edited by F. Brezzi, VII, 333 pages. 1985.

Vol. 1128: J. Elschner, Singular Ordinary Differential Operators and Pseudodifferential Equations. 200 pages. 1985.

Vol. 1129: Numerical Analysis, Lancaster 1984. Proceedings. Edited by P.R. Turner. XIV, 179 pages. 1985.

Vol. 1130: Methods in Mathematical Logic. Proceedings, 1983. Edited by C.A. Di Prisco. VII, 407 pages. 1985.

Vol. 1131: K. Sundaresan, S. Swaminathan, Geometry and Nonlinear Analysis in Banach Spaces. III, 116 pages. 1985.

Vol. 1132: Operator Algebras and their Connections with Topology and Ergodic Theory. Proceedings, 1983. Edited by H. Araki, C.C. Moore, Ş. Strătilă and C. Voiculescu. VI, 594 pages. 1985.

Vol. 1133: K.C. Kiwiel, Methods of Descent for Nondifferentiable Optimization. VI, 362 pages. 1985.

Vol. 1134: G.P. Galdi, S. Rionero, Weighted Energy Methods in Fluid Dynamics and Elasticity. VII, 126 pages. 1985.

Vol. 1135: Number Theory, New York 1983–84. Seminar. Edited by D.V. Chudnovsky, G.V. Chudnovsky, H. Cohn and M.B. Nathanson. V, 283 pages. 1985.

Vol. 1136: Quantum Probability and Applications II. Proceedings, 1984. Edited by L. Accardi and W. von Waldenfels. VI, 534 pages. 1985.

Vol. 1137: Xiao G., Surfaces fibrées en courbes de genre deux. IX, 103 pages. 1985.

Vol. 1138: A. Ocneanu, Actions of Discrete Amenable Groups on von Neumann Algebras. V, 115 pages. 1985.

Vol. 1139: Differential Geometric Methods in Mathematical Physics. Proceedings, 1983. Edited by H. D. Doebner and J. D. Hennig. VI, 337 pages. 1985.

Vol. 1140: S. Donkin, Rational Representations of Algebraic Groups. VII, 254 pages. 1985.

Vol. 1141: Recursion Theory Week. Proceedings, 1984. Edited by H.-D. Ebbinghaus, G.H. Müller and G.E. Sacks. IX, 418 pages. 1985.

Vol. 1142: Orders and their Applications. Proceedings, 1984. Edited by I. Reiner and K. W. Roggenkamp. X, 306 pages. 1985.

Vol. 1143: A. Krieg, Modular Forms on Half-Spaces of Quaternions. XIII, 203 pages. 1985.

Vol. 1144: Knot Theory and Manifolds. Proceedings, 1983. Edited by D. Rolfsen. V, 163 pages. 1985.

Vol. 1145: G. Winkler, Choquet Order and Simplices. VI, 143 pages. 1985.

Vol. 1146: Séminaire d'Algèbre Paul Dubreil et Marie-Paule Malliavin. Proceedings, 1983–1984. Edité par M.-P. Malliavin. IV, 420 pages. 1985.

Vol. 1147: M. Wschebor, Surfaces Aléatoires. VII, 111 pages. 1985.

Vol. 1148: Mark A. Kon, Probability Distributions in Quantum Statistical Mechanics. V, 121 pages. 1985.

Vol. 1149: Universal Algebra and Lattice Theory. Proceedings, 1984. Edited by S. D. Comer. VI, 282 pages. 1985.

Vol. 1150: B. Kawohl, Rearrangements and Convexity of Level Sets in PDE. V, 136 pages. 1985.

Vol 1151: Ordinary and Partial Differential Equations. Proceedings, 1984. Edited by B.D. Sleeman and R.J. Jarvis. XIV, 357 pages. 1985.

Vol. 1152: H. Widom, Asymptotic Expansions for Pseudodifferential Operators on Bounded Domains. V, 150 pages. 1985.

Vol. 1153: Probability in Banach Spaces V. Proceedings, 1984. Edited by A. Beck, R. Dudley, M. Hahn, J. Kuelbs and M. Marcus. VI, 457 pages. 1985.

Vol. 1154: D.S. Naidu, A.K. Rao, Singular Pertubation Analysis of Discrete Control Systems. IX, 195 pages. 1985.

Vol. 1155: Stability Problems for Stochastic Models. Proceedings, 1984. Edited by V.V. Kalashnikov and V.M. Zolotarev. VI, 447 pages. 1985.

Vol. 1156: Global Differential Geometry and Global Analysis 1984. Proceedings, 1984. Edited by D. Ferus, R.B. Gardner, S. Helgason and U. Simon. V, 339 pages. 1985.

Vol. 1157: H. Levine, Classifying Immersions into $\mathbb{R}^4$ over Stable Maps of 3-Manifolds into $\mathbb{R}^2$. V, 163 pages. 1985.

Vol. 1158: Stochastic Processes – Mathematics and Physics. Proceedings, 1984. Edited by S. Albeverio, Ph. Blanchard and L. Streit. VI, 230 pages. 1986.

Vol. 1159: Schrödinger Operators, Como 1984. Seminar. Edited by S. Graffi. VIII, 272 pages. 1986.

Vol. 1160: J.-C. van der Meer, The Hamiltonian Hopf Bifurcation. VI, 115 pages. 1985.

Vol. 1161: Harmonic Mappings and Minimal Immersions, Montecatini 1984. Seminar. Edited by E. Giusti. VII, 285 pages. 1985.

Vol. 1162: S.J.L. van Eijndhoven, J. de Graaf, Trajectory Spaces, Generalized Functions and Unbounded Operators. IV, 272 pages. 1985.

Vol. 1163: Iteration Theory and its Functional Equations. Proceedings, 1984. Edited by R. Liedl, L. Reich and Gy. Targonski. VIII, 231 pages. 1985.

Vol. 1164: M. Meschiari, J.H. Rawnsley, S. Salamon, Geometry Seminar "Luigi Bianchi" II – 1984. Edited by E. Vesentini. VI, 224 pages. 1985.

Vol. 1165: Seminar on Deformations. Proceedings, 1982/84. Edited by J. Ławrynowicz. IX, 331 pages. 1985.

Vol. 1166: Banach Spaces. Proceedings, 1984. Edited by N. Kalton and E. Saab. VI, 199 pages. 1985.

Vol. 1167: Geometry and Topology. Proceedings, 1983–84. Edited by J. Alexander and J. Harer. VI, 292 pages. 1985.

Vol. 1168: S.S. Agaian, Hadamard Matrices and their Applications. III, 227 pages. 1985.

Vol. 1169: W.A. Light, E.W. Cheney, Approximation Theory in Tensor Product Spaces. VII, 157 pages. 1985.

Vol. 1170: B.S. Thomson, Real Functions. VII, 229 pages. 1985.

Vol. 1171: Polynômes Orthogonaux et Applications. Proceedings, 1984. Edité par C. Brezinski, A. Draux, A.P. Magnus, P. Maroni et A. Ronveaux. XXXVII, 584 pages. 1985.

Vol. 1172: Algebraic Topology, Göttingen 1984. Proceedings. Edited by L. Smith. VI, 209 pages. 1985.

Vol. 1173: H. Delfs, M. Knebusch, Locally Semialgebraic Spaces. XVI, 329 pages. 1985.

Vol. 1174: Categories in Continuum Physics, Buffalo 1982. Seminar. Edited by F.W. Lawvere and S.H. Schanuel. V, 126 pages. 1986.

Vol. 1175: K. Mathiak, Valuations of Skew Fields and Projective Hjelmslev Spaces. VII, 116pages. 1986.

Vol. 1176: R.R. Bruner, J.P. May, J.E. McClure, M. Steinberger, H_∞ Ring Spectra and their Applications. VII, 388 pages. 1986.

Vol. 1177: Representation Theory I. Finite Dimensional Algebras. Proceedings, 1984. Edited by V. Dlab, P. Gabriel and G. Michler. XV, 340 pages. 1986.

Vol. 1178: Representation Theory II. Groups and Orders. Proceedings, 1984. Edited by V. Dlab, P. Gabriel and G. Michler. XV, 370 pages. 1986.

Vol. 1179: Shi J.-Y. The Kazhdan-Lusztig Cells in Certain Affine Weyl Groups. X, 307 pages. 1986.

Vol. 1180: R. Carmona, H. Kesten, J.B. Walsh, École d'Été de Probabilités de Saint-Flour XIV – 1984. Édité par P.L. Hennequin. X, 438 pages. 1986.

Vol. 1181: Buildings and the Geometry of Diagrams, Como 1984. Seminar. Edited by L. Rosati. VII, 277 pages. 1986.

Vol. 1182: S. Shelah, Around Classification Theory of Models. VII, 279 pages. 1986.

Vol. 1183: Algebra, Algebraic Topology and their Interactions. Proceedings, 1983. Edited by J.-E. Roos. XI, 396 pages. 1986.

Vol. 1184: W. Arendt, A. Grabosch, G. Greiner, U. Groh, H.P. Lotz, U. Moustakas, R. Nagel, F. Neubrander, U. Schlotterbeck, One-parameter Semigroups of Positive Operators. Edited by R. Nagel. X, 460 pages. 1986.

Vol. 1185: Group Theory, Beijing 1984. Proceedings. Edited by Tuan H.F. V, 403 pages. 1986.

Vol. 1186: Lyapunov Exponents. Proceedings, 1984. Edited by L. Arnold and V. Wihstutz. VI, 374 pages. 1986.

Vol. 1187: Y. Diers, Categories of Boolean Sheaves of Simple Algebras. VI, 168 pages. 1986.

Vol. 1188: Fonctions de Plusieurs Variables Complexes V. Séminaire, 1979–85. Edité par François Norguet. VI, 306 pages. 1986.

Vol. 1189: J. Lukeš, J. Malý, L. Zajíček, Fine Topology Methods in Real Analysis and Potential Theory. X, 472 pages. 1986.

Vol. 1190: Optimization and Related Fields. Proceedings, 1984. Edited by R. Conti, E. De Giorgi and F. Giannessi. VIII, 419 pages. 1986.

Vol. 1191: A.R. Its, V.Yu. Novokshenov, The Isomonodromic Deformation Method in the Theory of Painlevé Equations. IV, 313 pages. 1986.

Vol. 1192: Equadiff 6. Proceedings, 1985. Edited by J. Vosmansky and M. Zlámal. XXIII, 404 pages. 1986.

Vol. 1193: Geometrical and Statistical Aspects of Probability in Banach Spaces. Proceedings, 1985. Edited by X. Fernique, B. Heinkel, M.B. Marcus and P.A. Meyer. IV, 128 pages. 1986.

Vol. 1194: Complex Analysis and Algebraic Geometry. Proceedings, 1985. Edited by H. Grauert. VI, 235 pages. 1986.

Vol.1195: J.M. Barbosa, A.G. Colares, Minimal Surfaces in $\mathbb{R}^3$. X, 124 pages. 1986.

Vol. 1196: E. Casas-Alvero, S. Xambó-Descamps, The Enumerative Theory of Conics after Halphen. IX, 130 pages. 1986.

Vol. 1197: Ring Theory. Proceedings, 1985. Edited by F.M.J. van Oystaeyen. V, 231 pages. 1986.

Vol. 1198: Séminaire d'Analyse, P. Lelong – P. Dolbeault – H. Skoda. Seminar 1983/84. X, 260 pages. 1986.

Vol. 1199: Analytic Theory of Continued Fractions II. Proceedings, 1985. Edited by W.J. Thron. VI, 299 pages. 1986.

Vol. 1200: V.D. Milman, G. Schechtman, Asymptotic Theory of Finite Dimensional Normed Spaces. With an Appendix by M. Gromov. VIII, 156 pages. 1986.